10-MINUTE MATHS MIND-STRETCHERS

SOUTH SCHOOL
WICK

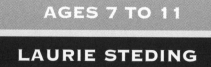

AGES 7 TO 11

LAURIE STEDING

Adapted from *Ten-minute Math Mind-stretchers* © 1997, Laurie Steding, published by Scholastic Inc, New York, USA.

Editor
Jane Gartside

Assistant Editor
David Sandford

Designer
Heather Sanneh

Illustrations
Garry Davies

Cover artwork
Geo Parkin

This edition © 2003 Scholastic Ltd

Designed using Adobe Pagemaker

Published by Scholastic Ltd,
Villiers House,
Clarendon Avenue,
Leamington Spa,
Warwickshire CV32 5PR

Printed by Bell & Bain Ltd, Glasgow

3 4 5 6 7 8 9 0   4 5 6 7 8 9 0 1 2

British Library Cataloguing-in-Publication Data
A catalogue record for this book is available from the British Library.

ISBN 0-439-98359-2

CONTENTS

INTRODUCTION

The rapidly changing world of maths education means there is so much information to share with children and so little time to fit more into our already over-committed days. That's why this book was written – to give children a ten-minute taste of maths problems, to make them think hard for a short period of time and to stimulate their curiosity about things they may not previously have thought of as maths concepts at all.

One of the main challenges for teachers today is the need to simultaneously address the needs of children of differing abilities – stretching gifted children while supporting less able children. For this reason, the problems and activities in *Ten-minute Maths* are varied in their level of difficulty. Some may seem easy and will give children opportunities to demonstrate their strengths; others may seem hard and will get children exploring and practising skills that need strengthening. Since children learn well from each other, many of the problems and activities are designed to be done with a partner or in small groups.

The problems in this book correspond to the *National Numeracy Strategy: Framework for Teaching Mathematics* (NNS) and are designed to supplement a well-rounded maths curriculum. A sample activity is shown overleaf to show you what each problem includes.

The NNS organises the maths curriculum into five broad 'strands': Numbers and the number system; Calculations; Solving problems; Handling data; and Measures, shapes and space. Inevitably there is crossover between strands in the problems, but the curriculum reference given indicates the main teaching objective that the problem explores.

Each chapter also includes *'Across the curriculum'* sections, which describe more elaborate cross-curricular activities designed to help the children realise that maths is everywhere around them.

These problems take only a short period of time to do but it is important that time is spent after each one to discuss the reasoning and elicit the strategies children used to arrive at their answers. Providing an environment in which children discuss, demonstrate and justify their mathematical thinking will encourage them to take some intellectual risks and expose them to new ways of doing maths.

The problems and activities in *Ten-minute Maths* can also be used for homework. In this case it is still important to ensure that the answers are discussed in class.

Enjoy using this book to stretch your children's minds a bit each day and to help them see that maths is fun!

How the activities are organised

An icon indicating how the activity is to be organised.

The National Numeracy Strategy curriculum reference.

The answer, which appears below the problem.

INDIVIDUAL 3

Rectangles for the number 24

Rapid recall of × and ÷ facts

Write the multiplication sentence that describes the rectangle in figure 2 on photocopiable page 42. (First write the number of squares across, then write the number of squares down.)

Cut out the squares. Rearrange them to form as many different rectangles for the number 24 as you can. Write a multiplication sentence for each one. Take some squares away and experiment with making rectangles for lesser numbers. Find a number for which you can make six rectangles; then five rectangles.

The multiplication sentence that describes the rectangle in figure 2 is 4 × 6 = 24. Eight different rectangles can be made from the 24 squares. The multiplication sentences representing them are: 1 × 24 = 24, 24 × 1 = 24, 2 × 12 = 24, 12 × 2 = 24, 3 × 8 = 24, 8 × 3 = 24, 4 × 6 = 24, 6 × 4 = 24. 20, 18, and 12 can each be represented by six different rectangles, and 16 by five rectangles.

INDIVIDUAL 4

Identifying prime numbers

Properties of number

You could make only two rectangles with five squares: a 1 × 5 rectangle and a 5 × 1 rectangle. The number 5 is a prime number. Each prime number has only two factors, itself and 1. Make a list of all the prime numbers between 1 and 20. You can use the squares from figure 2 (photocopiable page 42) to help.

Prime numbers between 1 and 20: 2, 3, 5, 7, 11, 13, 17 and 19.

INDIVIDUAL 5

Identifying square numbers

Properties of number

If you had 25 squares, you could make one large square (5 × 5 squares). The number 25 is a square number. Use the squares from the rectangle in figure 2 (photocopiable page 42) to find a number between 12 and 20 that is a square number. Write a multiplication sentence to describe the large square you made.

Find two square numbers that are less than 10. Write a multiplication sentence for each one. What do you notice about the multiplication sentences you wrote for the square numbers?

The number 16: 4 × 4 = 16; the numbers 4 (2 × 2 = 4) and 9 (3 × 3 = 9). Children should notice that the two factors in each multiplication sentence are the same.

INDIVIDUAL 6

Square numbers between 1 and 144

Properties of number

Use the 'Table square' (photocopiable page 43), and colour in the product you get when you multiply each number from 1 to 12 by itself. For example, 1 × 1 = 1, so colour in the number 1 in the 1 column; 2 × 2 = 4, so colour in the number 4 in the 2 column, and so on. Remember that when two factors are the same number, the product is a square number. What pattern do you see when you colour in the square numbers in the chart?

The square numbers form a diagonal line from the top left of the chart to the bottom right.

Chapter One

Solving REAL-LIFE problems

Maths on the job

Reasoning and generalising about number

Working in pairs, ask children to list five ways that a pizza delivery driver uses maths in his or her job. Share their ideas in a class discussion.

Answers will vary, but may include: calculating time, distance, and cost; finding addresses; giving change; keeping track of hours worked; packaging food.

Maths in sports

Reasoning and generalising about number

List three ways a member of a swimming team uses maths. Compare your partner's list with your own and see how many different ways you have found between you.

Answers will vary, but may include: timing races, calculating speed, comparing scores, calculating scores, determining depth. The total number of ways will vary from three to six.

Maths in weather forecasts

Measures

What maths vocabulary does a weather forecaster use when he or she talks about the weather? Make a list. Listen to a radio or TV weather report and add to your list any maths words the forecaster uses that aren't already on your list.

Lists will vary, but may include: words relating to temperature – degrees Celsius; time of weather changes or duration of storms – hours, overnight; wind speeds – kilometres or miles per hour; predicted or actual measures of rain – millimetres, centimetres or inches.

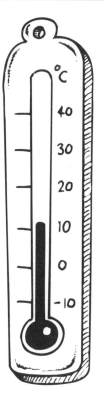

4 Comparing and ordering sports scores

Place value and ordering

Working in pairs, ask children to write down five different sports and make up possible match scores for each. Then ask them to write the sports in order from the greatest number of points in a match to the least. Children should then share what they wrote.

Sports and scores will vary. Possible sports and their score order: cricket, snooker, rugby, netball, football.

5 Making maths sense

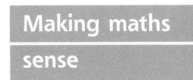

Problems involving 'real life' and measures

William likes to exaggerate when he talks. Work with your group to rewrite each exaggeration so that it is believable.

1   *I was on a plane that flew the length of the country in 15 minutes!*

2   *I only need one hour of sleep every night!*

3   *Last night 100cm of rain fell in my garden!*

Answers will vary, but possible responses may include: 1 I was on a plane that flew the length of the country in two hours; 2 I only need seven hours of sleep every night; 3 Last night 2cm of rain fell in my garden.

6 Making maths sense

Problems involving 'real life' and measures

A new announcer on the radio made this report: 'A bank robber got away from the First National Bank with 5000 pence. He escaped in a car which sped away at 70 metres per hour. Police set up a roadblock on the high street and captured the thief. Now, we'll take a commercial break and continue this story in one year.' Which three words in the story do not make sense? Write down the words you would use so that the story makes sense.

Change pence to pounds; change metres to kilometres; change year to minute.

7 Comparing ways to travel

Problems involving 'real life'

Sally, Tania, Marcus, Tim, Sarah and James are having a race to find out who can get around the block the fastest. They each move in a different way: Sally saunters, Tania trots, Marcus marches, Tim tiptoes, Sarah strides, and James jogs. Write the children's names in the order you think they will finish the race. Compare your answers with your partner's and discuss why you wrote them in the order you did.

Accept reasonable answers. Trotting and jogging are generally faster than striding and marching, which in turn, are faster than sauntering or tiptoeing. One possible response: Tania, James, Sarah, Marcus, Sally Tim.

Making a timetable

Handling data

'Make sure you allow yourself enough time so that you're not late for school today,' said Rebecca's mum. Work with your partner to list the things Rebecca probably needs to do in the morning before school. Beside each item on your list, write the amount of time you think it should take. How much time will it take her to do all the things on the list? If she needs to leave for school at 8:00 to be on time, what is the latest time Rebecca can get up?

Lists will vary, but may include: eat, wash, get dressed and collect homework. Times will vary, but the time it takes her to do everything on the list, when added to the time she needs to get up, should bring the time to 8:00.

Completing a word problem

Problems involving 'real life' and money.
Write a question to complete this word problem:

Angie and her sister went to the shopping centre to buy hats. Angie bought a hat that cost £4.50. Her sister bought a hat on sale for £3.95.

Swap problems with your partner and solve them. Check each other's work.

Questions will vary, but may include: How much did the two hats cost altogether? or How much more did Angie pay for her hat than her sister paid?

ACROSS THE CURRICULUM

1 Decorate the words on each copy of the 'Maths vocabulary sheet' and cut the words out.
2 Glue two words that match from each sheet back-to-back, with pieces of string or wool in between, hanging two or three words on each string.
3 Tie the strings to the stick or coat-hanger, placing them so the mobile balances.

LITERACY

Maths words scavenger hunt
Distribute copies of the 'Maths vocabulary sheet' on page 15 to small groups or individuals. Instruct the children to:

D & T

Each child needs two copies of the 'Maths vocabulary sheet' on page 15, crayons or felt-tips, a stick or coat-hanger, scissors, string, and glue. Discuss what each of the vocabulary words means and ask children to give examples.
Give children the following instructions to make the mobiles:

NB: you may want to omit or add vocabulary words so that the mobiles more closely reflect children's vocabulary levels.

1 Look for the words on the sheet in maths books, newspapers, magazines, on TV, or on signs around their neighbourhood.
2 Record where each one is found and copy the sentence in which it is used. Allow children a few days to look for the words. Set aside some time for them to share their findings.

Calculating costs

Problems involving 'real life' and money.
'The wind knocked part of my fence down,' groaned Ms Glenn, 'and it's going to cost a lot of money to fix it.' What does Ms Glenn need to know before she can work out how much it will cost to repair the fence?

Accept reasonable responses, such as: she needs to know the length and height of the section of fence that needs to be repaired, how much each metre of fencing costs, and how much it will cost to pay somebody to put in the new fencing if she cannot do it herself.

Measuring time

Problems involving 'real life' and money.
List these situations in order, from the shortest to longest amount of time it seems you spend waiting:

★ for food in a restaurant

★ for a birthday or special holiday

★ for a bus

★ your turn in a game.

Compare your list with your partner's. What is the same about your lists? What is different?

Lists will vary.

Comparing sale prices

Problems involving 'real life' and money.
Anna went to the shopping centre to buy a pair of jeans. She found one pair that was on sale with $\frac{1}{3}$ off the normal price, and another pair that was on sale with $\frac{1}{4}$ off the normal price. Write down the steps that Anna needs to follow to find which pair of jeans is cheapest.

Answers may vary. One possible answer: Find out the normal price of each pair of jeans. Divide the pair that has $\frac{1}{4}$ off by four, then subtract that amount from the normal price. Divide the pair that has $\frac{1}{3}$ off by three, then subtract that amount from the normal price. Compare the two sale prices.

Using coupons to save

Problems involving 'real life' and money.
Daniel's mum says that if he cuts out coupons for things they need to buy, and helps her shop, he can keep all the money saved by using the coupons. List five things that they might buy using coupons. How can Daniel keep track of how much money is being saved?

Lists may vary. Ways Daniel can keep track might include looking at the supermarket receipt, identifying the coupon deductions, and finding the sum of all the deductions.

Numerical prefixes

Properties of number
How are these words related?

triple   tripod   triplets   tricycle   trio   triathlon

What are some words that mean 'two of' something?
Draw a picture for each word.

They all mean 'three of' something. Words for 'two of' may vary, but could include: pair, bicycle, biathlon, double, twins; Check the children's drawings.

ACROSS THE CURRICULUM

CITIZENSHIP

Maths work
Ask children to interview relatives and neighbours to find out how they use maths in their jobs. Make a class chart of the information they gather.

Days, months and seasons

Measures
Drifting on a raft in the middle of the ocean, a man spent over 120 days alone before being rescued!

Work with your partner. About how many months is 120 days? What season was it 120 days ago? List four things that you have done or that have happened to you during the last 120 days.

---

120 days is about 4 months. The season will vary; lists will vary.

Venn diagrams 1

Handling data
Make a Venn diagram with two intersecting circles. Label one circle 'Me' and the other circle with the name of a member of your family. In the circle labelled 'Me', list characteristics that describe you, but not your family member. In the other circle, list characteristics that describe your family member, but not you. In the space where the circles intersect, list characteristics you both share.

Venn diagrams will vary.

Venn diagrams 2

Handling data
Make a Venn diagram with three intersecting circles. Label the circles 'Bumpy', 'Round', and 'Small'. Write the following words in the circle or intersection where they belong. Remember, some objects may not belong in any circle.

orange   marble   clock   dice   apple

biscuit   tree   kitchen roll   dictionary

bulletin board   drawing pin

Venn diagrams may vary. Accept reasonable responses.

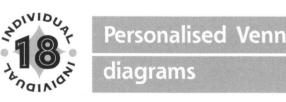

Personalised Venn diagrams

Handling data
Make another Venn diagram. Decide on the categories, such as 'Dogs' and 'Cats', and label the circles. Write characteristics in the circles such as *meows* for cats and *barks* for dogs. In the area where the circles intersect write shared characteristics such as *four-legged*. Show your Venn diagram to the members of your group and see if they can add any characteristics to it.

Check the children's Venn diagrams.

Logical thinking

Problems involving 'real life'

Everyone in the Jones family has an opinion about what to have for dessert. The choices are: ice cream, chocolate cake, apple pie, and oatmeal biscuits. Read

what each family member says and choose the one dessert they can agree on eating.

Cathy says, 'I don't want to eat anything cold.'

Liz says, 'I've eaten all the fruit I need to today.'

Derek says, 'I shouldn't eat chocolate.'

Faith says, 'I'm allergic to cake.'

The dessert they can all have is oatmeal biscuits.

Following

Properties and number sequences

Ask children to draw pictures according to the following instructions:

If you are more than eight years old, draw a Sun high up in the sky.

If you are eight years old or younger, draw a Sun low in the sky.

If you are a boy, draw several butterflies on your picture.

If you are a girl, draw several birds on your picture.

If your birthday is on an even-numbered day, draw flowers.

If your birthday is on an odd-numbered day, draw trees.

If you have more than four people in your family, draw fish in a pond.

If you have four people or fewer in your family, draw rabbits in the grass.

Ask children to add details to their pictures to complete them. Display all the finished pictures and call children out in turns to describe what they think the pictures reveal about the class as a whole and as individuals.

Writing directions

Properties of number

Working in small groups, challenge children to write a set of instructions for drawing pictures. Suggest that they use four or more categories, such as hair colour, the length of their names, whether or not they have a pet, whether or not they like to eat pizza, and so on. Each group should swap instructions and draw pictures accordingly. The groups should then check each other's drawings.

Answers will vary.

Estimating quantity

Estimating

Dried beans (or paper clips or coloured beads for example) are needed for this exercise. Ask children to estimate how many of each they think they can scoop in one handful. These estimates should be recorded. The children then count the actual number of beans they can scoop up. Ask how close their estimates were to the actual amount they can hold.

Answers will vary.

23 Making and adjusting estimates

Estimating and rounding
Choose a book without many pictures on the pages. Open the book and look at one of the pages.

★ *Write an estimate of how many words you think are on the page.*

★ *Count the number of words on the first line of the page, and the number of lines on the page. Estimate the total number of words again, using this information.*

★ *Is this estimate different from your first estimate?*

★ *Now count the words to find the actual number of words on the page. Which estimate was closest?*

Estimates will vary; the number of words on pages will vary. Either estimate may be closer to the actual number, but it is most likely that the second estimate will be closer, since it is based on more information.)

24 Finding needed information

Estimating
What information would you need to have in order to make a good estimate of the number of school lunches that will be eaten at school today?

Answers may vary. Possible answers include: the number of people who usually eat lunch, the menu, the day's attendance figure.

ACROSS THE CURRICULUM

MATHS

Find the secret number
Children should play this game in small groups. Each group needs a red, yellow and green crayon or marker. One player writes a number and the other players try to guess it, using clues from the number-writer.

How to play:
1 *Player A writes a three-digit number, keeping it hidden from the other players.*

2 *Player B writes down a guess.*

3 *Player A responds to the guess by drawing three dots, colour-coded as follows:*
★ *a yellow dot means that one of the digits is correct, but is in the wrong position*
★ *a green dot means that one of the digits is correct, and is in the correct position*
★ *a red dot means that neither the digit nor the position is correct. (It's important to stress to the children that the order of the dots does not correspond to the order of the digits in the number.)*

4 *Player C then uses this information to write down another guess.*

5 *Play continues with Players B and C guessing and Player A responding until the correct number is guessed.*

For example:
Player B guesses the number 356. Player A draws a red dot, a green dot, and a yellow dot to indicate that one digit is not correct in either number or position, one digit is correct in both number and position, and one digit is correct in number, but not in position. Player C guesses the number 362; player A draws a red dot, a green dot, and a green dot to indicate that one digit is not correct in either number or position, while two digits are correct in both number and position. Player D guesses the number 367. Player A tells them it's the correct number.

25 Things that come in dozens

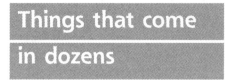

Problems involving 'real life'

In groups of two or three, ask children to brainstorm things that are grouped or packaged in sets of 12 and to draw or list them. What do they notice about their findings?

Answers will vary, but may include: eggs, doughnuts, or inches. Children may notice most or all of the items in their lists are food items.

26 Weather forecasts

Problems involving 'real life'

What information do you use to help you predict what the day's weather will be like when you get up in the morning? Why is knowing the day's weather forecast helpful? Write a story about how you had to change your plans for the day based on a weather forecast.

Answers may vary, but may include weather forecasts on the radio, TV or in the newspaper; the look and feel of the weather when you look or step outside. Knowing the weather forecast helps you decide what to wear and what to do that day. Stories will vary.

ACROSS THE CURRICULUM

MATHS

Estimation station

Create an 'estimation station' in an area of the classroom. In it, display a see-through container filled with some items such as dried beans or blocks. Vary the item or quantity from day to day. Children can take turns being in charge of the estimation station and can bring in objects and containers from home. Each day ask children to write an estimate of how many objects are in the container. At the end of each day, count the objects together. Determine who gave the closest estimate. Encourage children to discuss the estimation strategies they used.

Suggested objects: pencils, cereal, dried pasta, marbles, pebbles, dried beans, seeds, shells, pennies, rice, peanuts, jelly beans.

Maths vocabulary

equivalent	more	under	capacity
fewer	less	over	increase
greater	total	between	average
sum	product	equal	range
difference	percent	value	about

📖 SCHOLASTIC

Chapter Two

ADDITION and SUBTRACTION

PAIRS · 1 · A fairy tale addition problem

Mental calculation strategies (+ and –)

How many animals would there be altogether if the bears (but not Goldilocks) invited the pigs (but not the wolf) and the Billy goats (but not the troll) over for a party? Draw a picture to show your answer. Write a problem of your own that uses numbers from stories. Swap problems with your partner. Draw a picture to show the answer.

---

3 + 3 + 3 = 9 animals. Check children's drawings.

INDIVIDUAL · 2 · Adding points

Mental calculation strategies (+ and –)

Suppose colours were assigned the following points:

purple = 6 points

red = 5 points

green = 4 points

yellow = 3 points

blue = 2 points

white = 1 point
(other colours are worth 0 points)

Look at your clothing. Count each colour only once in each piece of clothing. Add up the points. What is your colour point total?

Make up your own colour point system and add up your points again. How does your new total compare with your first total?

Totals will vary: colour point systems will vary. New totals can be greater, less or the same as the first totals.

Adding points

Mental calculation strategies (+ and −)

Suppose each letter of the alphabet is worth points. The letter A is worth 1 point, B is worth 2 points, C is worth 3 points and so on up to Z, which is worth 26 points. Make a table to show how many points each letter is worth.

Add up the letters in your name using the points in the table. Then add up the letters in your partner's name. Who has the greatest number of points in their name?

Susan

$19 + 21 + 19 + 1 + 14 = 74$

A=1	B=2	C=3	D=4	E=5	F=6	G=7	H=8
I=9	J=10	K=11	L=12	M=13	N=14	O=15	
P=16	Q=17	R=18	S=19	T=20	U=21		
V=22	W=23	X=24	Y=25	Z=26			

Totals will vary; first names with the greatest and least number of points will vary.

Illustrating and writing a word problem

Paper and pencil procedures (+ and −)

Think of an addition problem that has 14 as its sum. First draw a picture of the problem, then write it as a word problem.

Illustrations and problems will vary.

Add pairs to make 100

Paper and pencil procedures (+ and −)

Work with the other members of your group to write all the different combinations of two whole numbers that add up to 100. What strategies did you use to find them? Compare your strategies with another group's. Did your group use the same ones? Explain.

0 + 100, 1 + 99, 2 + 98, 3 + 97... 50 + 50. Strategies will vary. Possible strategies include: to begin with combinations of numbers ending in 0, such as 10 + 90, or finding a pattern such as 100 + 0, 99 + 1, 98 + 2, and so on; groups may or may not have used the same strategies.

Writing numbers

Pencil and paper procedures (+ and −)

Use the digits 2, 4, 5 and 8 to write two two-digit numbers whose sum is 100. Write two possible answers.

The two possible answers are 48 + 52 and 58 + 42.

ACROSS THE CURRICULUM

M A T H S

Addition stones

This is a Native American game. For playing pieces you will need four smooth stones. Draw each of the simple designs shown below on one side of each stone. The other side of each stone remains blank.

Any number of children can play. Players take turns placing the stones in a container, shaking the container and spilling the stones onto a hard surface.

They receive points for each design showing, according to the chart below, adding points for their total score. The player with the greatest score after three rounds wins.

Blank side of any stone	=	0 points
Moon	=	5 points
Star	=	10 points
Sun	=	15 points
Snowflake	=	20 points

Using mental maths

Mental calculation strategies (+ and −)
Use mental maths to find the answer to each question. Explain which strategy you used for each one.

1   *What number comes after 17 + 8?*

2   *Which is greater, 21 + 11 or 20 + 14?*

3   *What is 68 + 42?*

---

1 26; 2 20 + 14; 3 110; strategies may vary.

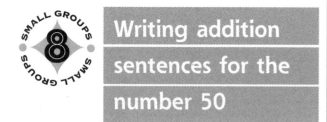

Writing addition sentences for the number 50

Paper and pencil procedures (+ and −)
Write ten addition sentences for the number 50. You can use as many numbers in each addition sentence as you like. For example, a sentence such as 10 + 10 + 10 + 20 = 50.

Compare your number sentences with other members of your group. How many different number sentences does your group have altogether?

Do you think there are still more addition sentences you could write for the number 50? Explain.

Number sentences will vary; group totals of number sentences will vary. Children should realise that there are many more addition sentences that can be written for the number 50.

Addition rules

Paper and pencil procedures (+ and −)
Copy and complete each sentence below to write four addition rules.

1   *The sum of an odd number and an odd number is always _____*

2   *The sum of an even number and an even number is always _____*

3   *The sum of an odd number and an even number is always _____*

4 *If you change the order of the numbers, the sum _____ _____ _____*

Discuss your rules with your partner. Work together to write two examples of each rule.

---

1 The sum of an odd number and an odd number is always an even number; 2 the sum of an even number and an even number is always an even number; 3 the sum of an odd number and an even number is always an odd number; 4 if you change the order of the numbers, the sum remains the same. Examples will vary.

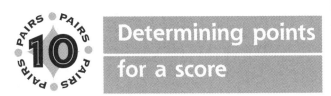

Determining points for a score

Mental calculation strategies (+ and –)
Look at the target on photocopiable page 23. In three rounds, throwing one dart each round, Harry got 37 points and Kim got 42 points. Write a number sentence to show how many points Harry got on each turn to get that total. Write a number sentence to show the number of points Kim got.

The order of numbers in number sentences may vary. Harry: 25 + 10 + 2 or 25 + 7 + 5; Kim: 25 + 10 + 7.

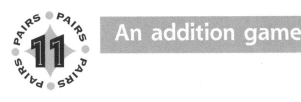

An addition game

Mental calculation strategies (+ and –)
Play a target game using the target on photocopiable page 23 and four small, light objects, like dry pasta. Place the target flat on a table, chair seat, or on the floor. Take turns with your partner, closing your eyes and dropping the pasta onto the target area. Add the points of the sections in which the pasta lands. The player with the greatest number of points after three rounds wins.

Scores will vary.

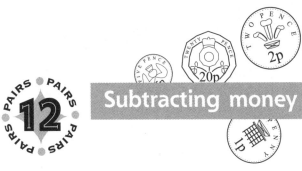

Subtracting money

Problems involving 'real life' and money
Sally's mum said that each time Sally didn't have her homework finished by tea time, 10p would be subtracted from her pocket money of £2.25 a week. At the end of the week, Sally received £1.85 pocket money. How many times was she late doing her homework?

Explain to your partner how you got your answer. Did you solve it the same way as your partner? Explain.

Four times. Solution methods may vary. Possible strategies include: count up in tens from £1.85 to £2.25, or write an equation – £2.25 – £1.85 = 40p, to find the amount subtracted, then determine the number of 10 pences in 40p.

Adding and subtracting money

Problems involving 'real life' and money
Grapes and strawberries cost £1. The grapes cost 10p less than the strawberries. Use mental maths to find the price of each item. Explain how you got your answer.

The grapes cost 45p and the strawberries cost 55p. Explanations may vary, but could include: use the 'guess and check' strategy. Start with both items priced at 50p. Subtract 10p from the strawberries to get the price of the grapes, which would be 40p; 50p + 40p = 90p, but 90p is too low. So add 5p to each price; 55p + 45p = £1.00.

ACROSS THE CURRICULUM

P·E

Exercise equations

Make a class list of the ten most popular exercise movements, such as touching toes or twisting at the waist. Specify a number of repetitions and assign points to each exercise. For example, deep knee bends (five times) – ten points. Display the list so everyone can see it.

With children working in small groups, create short (3–5 minute) routines consisting of exercises listed in the chart. Each group can teach the rest of the class its routine, and children can write number sentences to find out how many points the routine is worth. Ask children questions such as: 'How could you change the routine so the total number of points is ten less or ten more?'

14 Subtracting biscuits

INDIVIDUAL · INDIVIDUAL

Mental calculation strategies (+ and –)
Linda's aunt made two dozen biscuits. She told Linda and her cousins not to eat more than half of the biscuits. Linda ate four biscuits, Michael ate three biscuits, Caroline ate six biscuits and little Karon ate two biscuits. Did they eat more than half the biscuits? Explain.

Yes. Two dozen biscuits is 24. Half of 24 is 12. Linda and her cousins ate 4 + 3 + 6 + 2, or 15 biscuits. 15 is greater than 12.

15 Comparing statements

PAIRS · PAIRS · PAIRS · PAIRS

Mental calculation strategies (+ and –)
Without subtracting, compare these two statements:

$335 – 49$ $497 – 72$

Which expression has the greater number for an answer? How do you know? Discuss it with your partner. Did you both solve it the same way?

---

497 – 72 has the greater answer.
One possible response: look at the numbers in the tens places. In the first statement, three is less than four, so re-grouping is needed. The answer will have a two in the hundreds place. In the second statement, nine is greater than seven, so no re-grouping is needed. The answer to the second statement is greater than the answer to the first statement.

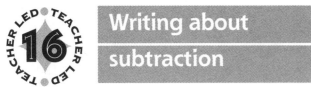 

16 Writing about subtraction

Pencil and paper procedures (+ and –)

Working in pairs, ask children to write news articles about their class. The article must include at least one subtraction situation. For example, children might write: 'On Thursday, 17 of Mrs Chatterjee's 30 Year 4 children brought their own lunches to school. The other 13 ate the school lunch of Macaroni Cheese.'

Compile the news articles into a class newspaper and give it a title such as *The Subtractions Gazette*.

News articles will vary.

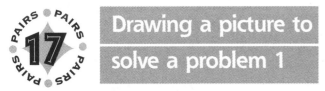

17 Drawing a picture to solve a problem 1

Problems involving 'real life'

Julie bought a bag of 28 marbles. When she got home she discovered there was a hole in the bag, and only nine marbles were left! How many marbles did Julie lose between the shop and home? Draw a picture of Julie's walk home from the shop, and show where all the missing marbles might be. Swap pictures with your partner and find all the missing marbles.

Julie lost 19 marbles; check children's drawings. There should be 19 marbles hidden in each picture.

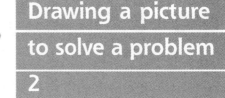

Problems involving 'real life'

The Crackley cricket team has 11 players. The team's uniform is red. At one game, seven players wore red socks. Three of the players with red socks also wore red sweatshirts. Draw a picture to answer the following questions:

1 *How many did not wear red socks or sweatshirts?*

2 *How many did not wear red socks?*

3 *How many did not wear red sweatshirts?*

Write a subtraction sentence for each answer.

Check children's drawings. 1 Four players were not wearing red socks or red sweatshirts: 11 – 7 = 4; 2 Four players were not wearing red socks: 11 – 7 = 4; 3 Eight players were not wearing red sweatshirts: 11 – 3 = 8.

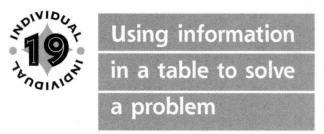

Handling data

Ms Kember's class is baking biscuits as a special treat. They need to decide what kind of biscuits to make. Of the 26 children in the class, three dislike chocolate chip biscuits and 12 dislike ginger biscuits; 18 children like biscuits with raisins and 20 like vanilla biscuits. Make a table showing how many kids like and dislike each kind of biscuit. What do you think the class should do?

Biscuit type	Number who like it	Number who don't like it
Chocolate chip	23	3
Biscuits with raisins	18	8
Vanilla biscuits	20	6
Ginger biscuits	14	12

Chart should resemble the one below. Decisions will vary, but may include: make two different kinds of biscuit; make chocolate chip biscuits, since the fewest number of children dislike them.

Pencil and paper procedures (+ and −)

Complete the 'Number buddies' chart (photocopiable page 24). Write two numbers on each T-shirt that, when added together, equal the number of that row. For example, in the row labelled '6' you can complete the number buddies to show 6 + 0, 5 + 1 and so on.

The order of expressions may vary, but each row should include some of the following:

Row 0: 0 + 0
Row 1: 1 + 0, 0 + 1
Row 2: 2 + 0, 0 + 2, 1 + 1
Row 3: 3 + 0, 0 + 3, 2 + 1, 1 + 2
Row 4: 4 + 0, 0 + 4, 3 + 1, 1 + 3, 2 + 2
Row 5: 5 + 0, 0 + 5, 4 + 1, 1 + 4, 3 + 2, 2 + 3
Row 6: 6 + 0, 0 + 6, 5 + 1, 1 + 5, 4 + 2, 2 + 4, 3 + 3
Row 7: 7 + 0, 0 + 7, 6 + 1, 1 + 6, 5 + 2, 2 + 5, 4 + 3, 3 + 4

ACROSS THE CURRICULUM

Marvellous machines

Explain to the children that they are going to invent machines! Ask them to work in small groups to brainstorm and decide what task their machine will perform. They should then draw the machine, making sure to include parts from the list below, and calculate the total cost of their machine. Each group can present its drawing to the class, explaining what the machine does and the cost. The class can create award categories (such as most interesting machine, least expensive machine, machine with the most parts, and so on), and vote for which group receives each award.

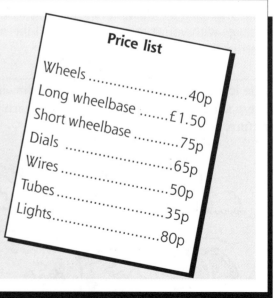

Price list

Wheels 40p
Long wheelbase £1.50
Short wheelbase 75p
Dials 65p
Wires 50p
Tubes 35p
Lights....................80p

Addition target

2

5

7

10

25

50

25

10

7

5

2

Number buddies

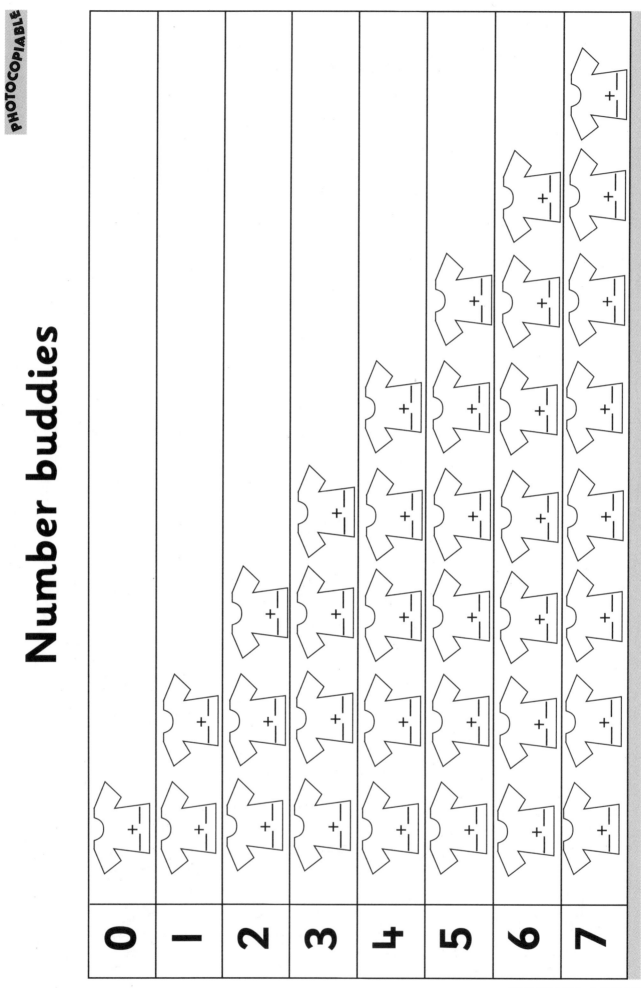

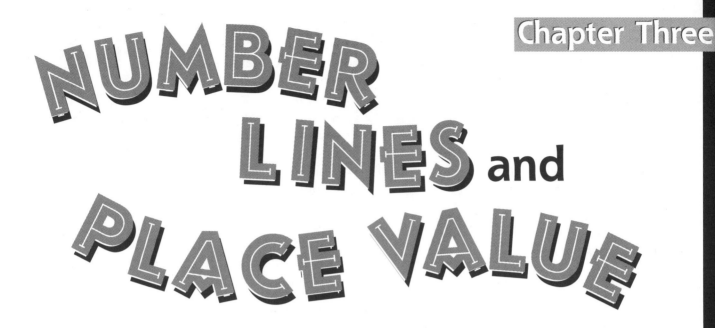

NUMBER LINES and PLACE VALUE

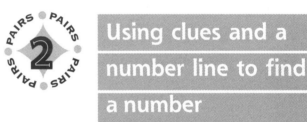

Visualising number lines

Place value, ordering and rounding

0, 1, 2, 3, 4, 5, 6, 7, 8, 9

Ask children to picture a number line in their minds. Ask them to start at zero, and mentally move along the number line as you count the following numbers: 1, 2, 3, 4, 5, 6, 7, 8, 9.

Ask them to go back to zero and to imagine themselves skipping over the even numbers and landing on the odd numbers. They should count aloud as they skip: '1, 3, 5, 7, 9, 11...' Ask them to continue 'skip-counting' until they reach the next odd number that:

★ *has the same digit in both the tens and the ones places;*

★ *is greater than 40, and has a number in the tens place that is 5 less than the number in the ones place.*

Continue giving children clues for numbers to find along their number lines. Encourage them to think of clues to give their classmates as well.

33; 49.

Using clues and a number line to find a number

Properties of number and number sequences
Picture a number line from 0 to 30 and then use these clues to find the mystery number:

★ *it is between 15 and 20*
★ *it's an odd number*
★ *it's closer to 15 than to 20.*

Write the number. Write some clues of your own for three different mystery numbers. Swap clues with your partner.

17; mystery number clues will vary.

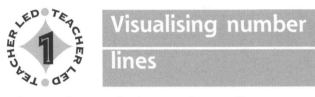

Chapter Three

Adding and subtracting on a number line

3 TEACHER LED · TEACHER LED

Place value, ordering and rounding

Ask children to picture a number line in their minds. Give them the following instructions:

1 *Start at 2.*

2 *Move ahead 5.*

3 *Go back 3.*

4 *Move ahead 4.*

Ask them to identify the number at which they've stopped and to write a number sentence describing how they moved along the number line.

They have stopped at 8; the number sentence is
2 + 5 – 3 + 4 = 8.

Describing numbers

4 SMALL GROUPS · SMALL GROUPS

Place value, ordering and rounding

Work with two or three other children, using the 'Numeral cards' from photocopiable page 30. Take turns choosing two cards and giving two clues as to what the numbers are. The clues can, for example, describe how far apart the numbers are on a number line, whether they are odd or even, what the sum would be if they were added together, and so on. A third clue can be given if the number is not guessed in one round.

For an additional challenge, choose three cards to make one two-digit number and one one-digit number, or four cards to make two two-digit numbers.

Numbers chosen and clues given will vary.

Writing greatest and least numbers

5 TEACHER LED · TEACHER LED

Place value, ordering and rounding

Use the 'Numeral cards' (photocopiable page 30) and take turns choosing three cards. The person who chose the cards uses the digits to write the greatest number possible. The other person uses the same digits to write the least number possible.

Continue until you've each written five numbers. Write all ten numbers in order from smallest to greatest.

Numbers will vary.

ACROSS THE CURRICULUM

On the number line (PE)
This game can be played with a large or small group of children. One player is the 'caller'. The other players pretend they are standing on number lines (so they need to leave plenty of space on either side of them). The caller identifies the number that players are to start on, then calls out another number and the players move the number of steps to the left or right necessary to put them on the new number. The caller continues, calling out other numbers for players to move to along their number lines. Players take turns being the caller.

You may also ask the caller to call out the number of steps to take, and in which direction, so players can say what number that puts them on.

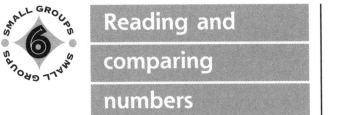

Reading and comparing numbers

Place value, ordering and rounding

Use the 'Numeral cards' (photocopiable page 30) to show your telephone number. Put it in the correct position on the 'Place value chart' (page 31) along with the telephone numbers of the other members of your group. Take turns reading your telephone numbers as whole numbers. Who has the phone number that forms the greatest number? The smallest number?

Numbers will vary.

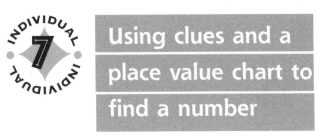

Using clues and a place value chart to find a number

Place value, ordering and rounding

Use the 'Numeral cards' and the 'Place value chart' (photocopiable pages 30 and 31), with these clues to find a mystery number:

★ the mystery number has three digits
★ the digit in the ones place is two less than the digit in the tens place
★ the digits in this mystery number add up to 5
★ the digit in the hundreds place is the same as the digit in the ones place.

Write the number.

The mystery number is 131.

Adding using a place value chart

Place value, ordering and rounding

Place each addition question below on the 'Place value chart' (photocopiable page 31) using the numeral cards from page 30. Then find the answer to each question.

1   2 + 4 = ?

2   12 + 4 = ?

3   32 + 4 = ?

4   102 + 4 = ?

5   2582 + 4 = ?

What pattern do you notice?

1 6; 2 16; 3 36; 4 106; 5 2586. The digit in the ones place doesn't change.

Estimating the number of words

Estimating

Work with two or three other children to find out roughly how many words there are on a page of a newspaper. Outline a small section of a newspaper page that is only words. Count each word in that section, then estimate how many sections of the same size are on the page. Multiply the number of words in the sample section by the number of sections on the page.

Compare your group's findings with those of another group. Describe another way to estimate the number of words on a newspaper page.

Estimates will vary. Other ways to estimate will vary: count the number of words in a line; count the number of lines on the page, and multiply the two numbers.

Money equivalencies

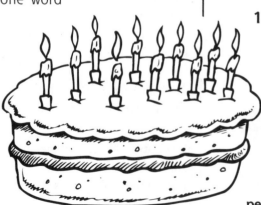

Problems involving 'real life' and money
Children at the Winlands Primary School decided to save pennies to buy something special for the school. They saved 10 000 pennies! How many pounds is that? Explain your answer in writing and with an illustration.

£100. Explanations may vary: since there are 100 pennies in £1, divide 10 000 by 100. The answer is 100. Check children's drawings.

11 INDIVIDUAL

Numbers of years

Measures
Tony's great-great-grandmother has her 100th birthday today. About how many days has she lived? How many decades are there in 100 years? (Look 'decade' up in a dictionary if you're not sure of its meaning.) What one word means 100 years?

About 36 500 days (100 × 365). There are 10 decades in 100 years; the word 'century' means 100 years.

12 PAIRS

Estimating number of breaths taken

Using a calculator
In pairs, take it in turns to time one minute, while also counting the number of breaths the other takes. Then switch roles. Estimate how many breaths you take in one day. About how many breaths do you take in one week? In one year? You may use a calculator to find the answers.

Number of breaths taken in one minute will vary. Estimates will vary. One possible answer, based on 20 breaths per minute: 20 × 60 = 1200 breaths each hour; 1200 × 24 = 28 800 breaths per day; 28 800 × 7 = 201 600 breaths per week; 201 600 × 52 = 10 483 200 breaths per year.

13 INDIVIDUAL

Calculating a salary

Using a calculator

Suppose a football player had just signed a contract for £100million, to be paid over the next five years.

1   *What would his salary be each year?*

2   About how much money is that each month? Each day?

You may need to use a calculator to solve this problem.

---

1 £20 000 000 per year; 2 About £1 666 667 per month; about £55 556 per day.

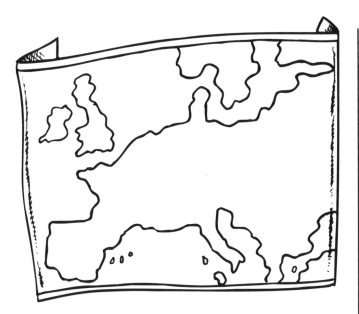

Maths superlatives

Ask children to bring in adverts and clippings from newspapers and magazines that include words such as *the most, the biggest, the longest* and so on. Can they identify what is being measured and what the measurement is? Discuss whether they think the claim is accurate.

LITERACY

14 Map skills

INDIVIDUAL

Measures

The Great Wall of China is over 3000 kilometres long. Look on a map of Europe, and use the scale to find two cities about that distance apart.

Answers will vary, but may include: London–Athens, Oslo–Istanbul, Barcelona–Helsinki, Dublin–Bucharest, Lisbon–Belgrade.

15 Distances in our solar system

INDIVIDUAL

Measures

Mercury is the closest planet to the Sun. It is 58 000 000 kilometres away from the Sun. Pluto, the furthest planet from the Sun, is 5 900 000 000 kilometres away from it. How far apart are Mercury and Pluto?

They are 5 842 000 000 kilometres apart.

Big events

LITERACY

Gather newspaper and magazine articles that use numbers in the thousands, hundreds of thousands, millions and beyond. Ask children to copy the numbers onto index cards. They can then put the cards in order from smallest to largest, practise reading the numbers, and match them with their corresponding articles.

Numeral cards

0	0	0	1	1
1	2	2	2	3
3	3	4	4	4
5	5	5	6	6
6	7	7	7	8
8	8	9	9	9

Place value chart

Millions	Hundred thousands	Ten thousands	Thousands	Hundreds	Tens	Ones

◣ S C H O L A S T I C

Chapter Four

MULTIPLICATION and DIVISION

1 Identifying multiplication and division situations

TEACHER LED • TEACHER LED • TEACHER LED

Problems involving 'real life'

In small groups, ask the children to try and brainstorm situations in their daily lives where they use multiplication and division. Write two columns headed 'Multiplication' and 'Division' on the board and record the examples given by each group as they share the results of their brainstorming session.

Examples will vary. Some possible examples: using division to determine the number of players to put on each team when playing a game; using multiplication to determine the price of three pencils when they know the price of one pencil.

2 Rectangles for the number 6

INDIVIDUAL • INDIVIDUAL

Rapid recall of × and ÷ facts

Look at the rectangle marked figure 1 on photocopiable page 42. The rectangle is one square across and six squares down, so we can write $1 \times 6 = 6$ to describe it. Turn the rectangle so it is six squares across and one square down. How does the multiplication sentence change?

Cut the squares out. Rearrange them to form as many other rectangles for the number six as you can. Write a multiplication sentence for each one. Take away one of the squares and arrange the five squares that are left into rectangles for the number five. How many rectangles can you make?

When the rectangle is turned it becomes a six-by-one rectangle, described by the multiplication sentence $6 \times 1 = 6$. The squares can be rearranged to show rectangles representing $2 \times 3 = 6$ and $3 \times 2 = 6$. Only two rectangles can be made with five squares: 5×1 and 1×5.

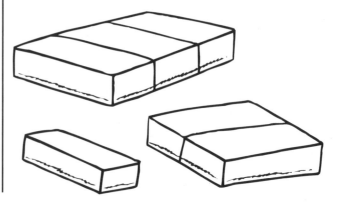

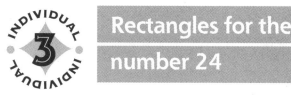

Rectangles for the number 24

Rapid recall of × and ÷ facts

Write the multiplication sentence that describes the rectangle in figure 2 on photocopiable page 42. (First write the number of squares across, then write the number of squares down.)

Cut out the squares. Rearrange them to form as many different rectangles for the number 24 as you can. Write a multiplication sentence for each one. Take some squares away and experiment with making rectangles for lesser numbers. Find a number for which you can make six rectangles; then five rectangles.

The multiplication sentence that describes the rectangle in figure 2 is 4 × 6 = 24. Eight different rectangles can be made from the 24 squares. The multiplication sentences representing them are: 1 × 24 = 24, 24 × 1 = 24, 2 × 12 = 24, 12 × 2 = 24, 3 × 8 = 24, 8 × 3 = 24, 4 × 6 = 24, 6 × 4 = 24. 20, 18, and 12 can each be represented by six different rectangles, and 16 by five rectangles.

Identifying prime numbers

Properties of number

You could make only two rectangles with five squares: a 1 × 5 rectangle and a 5 × 1 rectangle. The number 5 is a prime number. Each prime number has only two factors, itself and 1. Make a list of all the prime numbers between 1 and 20. You can use the squares from figure 2 (photocopiable page 42) to help.

Prime numbers between 1 and 20: 2, 3, 5, 7, 11, 13, 17 and 19.

Identifying square numbers

Properties of number

If you had 25 squares, you could make one large square (5 × 5 squares). The number 25 is a square number. Use the squares from the rectangle in figure 2 (photocopiable page 42) to find a number between 12 and 20 that is a square number. Write a multiplication sentence to describe the large square you made.

Find two square numbers that are less than 10. Write a multiplication sentence for each one. What do you notice about the multiplication sentences you wrote for the square numbers?

The number 16: 4 × 4 = 16; the numbers 4 (2 × 2 = 4) and 9 (3 × 3 = 9). Children should notice that the two factors in each multiplication sentence are the same.

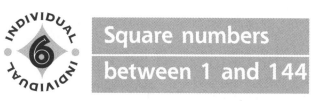

Square numbers between 1 and 144

Properties of number

Use the 'Table square' (photocopiable page 43), and colour in the product you get when you multiply each number from 1 to 12 by itself. For example, 1 × 1 = 1, so colour in the number 1 in the 1 column; 2 × 2 = 4, so colour in the number 4 in the 2 column, and so on. Remember that when two factors are the same number, the product is a square number. What pattern do you see when you colour in the square numbers in the chart?

The square numbers form a diagonal line from the top left of the chart to the bottom right.

7

Finding patterns of multiples

Properties of number and number sequences

Choose a number from the first column on the 'Table square' worksheet (photocopiable page 42). Colour in all the multiples of that number you find in the chart. For example, if you choose the number 7 from the first column, you would colour in the numbers 7, 14, 21 and so on, wherever they appear in the chart. Write a sentence about the pattern you see.

There are many different patterns for the children to find – for example, multiples of 2 make a crisscross pattern on the chart.

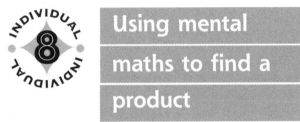 **8**

Using mental maths to find a product

Mental calculation strategies (× and ÷)

The Freeman family used five rolls of film taking photos on their holiday. Each roll of film had 24 pictures on it. Mrs Freeman wants to know how many photos that will be in all. She doesn't have a pencil and paper handy, nor a calculator. What mental calculations could she use to find the product?

Make up a problem similar to this one. Explain how you would solve it using a mental maths strategy.

Answers will vary. One possible answer: think of 24 as 20 + 4, multiply 20 by 5 and 5 by 4 and add the products: 100 + 20 = 120. The answer is 120 pictures. Problems and solutions will vary.

 **9**

Exploring the effects of doubling amounts

Properties of number and number sequences

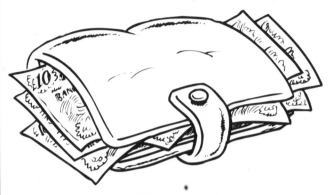

Suppose you were willing to take just one penny as pocket money this week, as long as each week after that the amount would double. For example, this week you would get 1p, next week 2p, the following week 4p, and so on. Do you think you'd be getting more or less than £10 for your pocket money by the twelfth week?

Make a table to show what your pocket money would be each week for 12 weeks. How much pocket money would you receive for the twelfth week? Is the amount greater or less than you thought it would be? How much pocket money would you have received in all?

Answers will vary. Pocket money would be as follows:
Week 1 = 1p
Week 2 = 2p
Week 3 = 4p
Week 4 = 8p
Week 5 = 16p
Week 6 = 32p
Week 7 = 64p
Week 8 = £1.28
Week 9 = £2.56
Week 10 = £5.12
Week 11 = £10.24
Week 12 = £20.48
You would receive £20.48 for the twelfth week. Answers may vary, but it is likely that this amount is more than the children thought it would be; the total amount of pocket money received is £40.95.

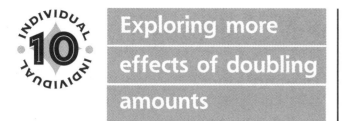

10 INDIVIDUAL — Exploring more effects of doubling amounts

Properties of number and number sequences

Suppose your pocket money was £3.00 a week. How much pocket money would you receive in 12 weeks? How does the total amount of pocket money you receive compare with the total amount of pocket money in problem 9?

£36.00; the amount is £4.95 less than the total in the previous problem.

11 INDIVIDUAL — Comparing smile statistics

Pencil and paper procedures (× and ÷)

The average person smiles about 15 times a day. How many times would that be in a week? A month? A year? Keep a tally to count the number of times you smile in one day. How does the number compare with the average?

---

15 × 7 = 105 times per week; 15 × 30 = 450 times per month; 15 × 365 = 5475 times per year. Tallies and comparisons will vary.

12 PAIRS — Solving a multi-step problem

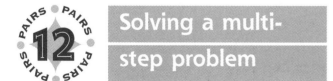

Pencil and paper procedures (× and ÷)

'Ooh,' moaned Mrs Barton, 'I have so many papers to mark!' Each child in her class had handed in 15 worksheets (except for two children who were absent). There are 29 children in her class. How many worksheets does Mrs Barton have to mark?

Work with a partner to write out the steps you take to solve this problem. Find another way to solve the problem by changing the order of the steps.

Mrs Barton has 405 worksheets to mark. Subtraction and multiplication are required to solve the problem, in either order. The subtraction can be done first and then the multiplication: 29 children – 2 absent children = 27; 27 children x 15 worksheets = 405; or the multiplication can be done first and then the subtraction.

13 PAIRS — Solving a multi-step problem

Pencil and paper procedures (× and ÷)

Work with a partner on this problem. Mrs Lee decided that each member of her family should eat three pieces of fresh fruit and two of fresh vegetables each day. There are four people in her family. She will buy apples, oranges, bananas, carrots, and potatoes this week. How many of each does she need to buy? How many fruits and vegetables is that in all? Explain what strategy you used to solve the problem.

She needs to buy 28 of each fruit and vegetable; a total of 28 x 5, or 140 in all. Strategies may vary. Children may find it helpful to make a chart similar to the one below.

Food item	Number needed each day	Number needed each week
Carrots	4	28
Potatoes	4	28
Apples	4	28
Oranges	4	28
Bananas	4	28
TOTAL	20	140

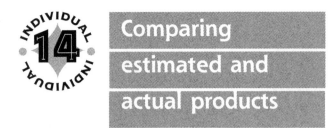

14 Comparing estimated and actual products

Mental calculation strategies (× and ÷)

1   Describe a way to multiply 80 x 20 using mental maths.

2   Round each factor to the nearest ten and estimate the products for each of the following: **a** 49 × 25 **b** 88 × 69 **c** 57 × 66

3   Is each estimate more or less than the actual product? How do you know?

---

1 One possible strategy: take off the two zeros and multiply 8 by 2 = 16, then add the two zeros back to make 1600; 2a 50 × 30 = 1500; b 90 × 70 = 6300; c 60 × 70 = 4200; 3 greater than; children should realise that they rounded each factor up so each estimate is greater than the actual products.

---

15 Draw a picture to solve a multiplication problem

Problems involving 'real life'

Draw a picture to solve this problem: on a street there are four houses. In each house there are six rooms. In each room there are five pieces of furniture. How many rooms did you draw? How many pieces of furniture? Compare your drawing with a friend's.

---

24 rooms:
4 × 6 = 24; 120
pieces of furniture:
24 × 5 = 120.

Favourite recipes

Ask children to bring in their favourite 'healthy snack' recipes. Children can work in small groups to increase or decrease the quantities of the ingredients in the recipes so that each recipe would make enough for the number of children in the class. Ask children to vote for one or two recipes to make. Enjoy the results!

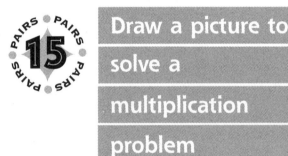

PSHE & SCIENCE

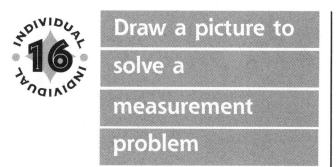

Draw a picture to solve a measurement problem

Measures

Amanda has a piece of paper 21cm wide and 18cm long. She wants to make a calendar. How many columns does she need? Most calendars have five rows, plus one more row for the names of the days of the week, making six rows. Amanda wants to make the columns and rows as large as possible. Draw a picture of the finished calendar, and label the length and width of one column and one row. What is the length and width of each box on the calendar?

Seven columns, one for each day of the week; the picture should show a calendar divided into 7 columns and 6 rows. One column should be labelled 18cm long and 3cm wide; one row should be 21cm long and 3cm wide; each box is 3cm long and 3cm wide.

Division with a fraction remainder

Mental calculation strategies (× and ÷)

Matt was proud of catching seven fish on a camping trip. The four other members of his family all wanted some fish when they smelled it cooking over the campfire. How can Matt divide the fish so everyone gets an equal share? How much will each person get?

One possible answer: each of the five family members gets one whole fish. The two remaining fish are each divided into 5 pieces, with each person receiving one-fifth of both fish. So each person gets a total of one and two-fifths fish.

Creating a pattern

INDIVIDUAL 18

Properties of number and number sequences

Kulwinder plans to make five beaded bracelets to give as gifts. She buys 50 white beads, 25 blue beads, and 25 red beads. She will use the same number of each colour bead for each bracelet. How many white beads will she use in one bracelet? How many blue beads? Red beads?

Draw or write a pattern to show how Kulwinder could string together the beads in one bracelet.

Ten white beads, five blue beads, five red beads; patterns will vary. One possible pattern: red-white-blue-white.

Working with remainders

SMALL GROUPS 19

Pencil and paper procedures (× and ÷)

Three classes of Year 5 children and their teachers are going on a field trip. Each group of five children needs an adult with them. They will invite some parents to help on the trip. Permission slips have been returned by 92 students. Each bus seats 40 people. With a partner, work out:

★ *how many adults will be needed*

★ *how many parent will be needed to help*

★ *how many buses will be needed.*

---

92 ÷ 5 = 18 remainder 2, so 19 adults are needed. There are already three teachers in the group, so 19 − 3 = 16 parents will be needed. The total number of people going on the trip is 92 + 3 + 16 = 111. One bus seats 40 people, two buses seat 80 people, three buses seat 120 people, so three buses will be needed.

Determining the best buy

INDIVIDUAL 20

Problems involving 'real life' and money

Francis was comparing different brands of coloured pencils. He saw one set of eight pencils for £1.99, another set of 12 pencils for £2.69, and a third set of ten pencils for £2.39. How much does a single pencil cost in the eight-pencil set? The 12 pencil set? The 10 pencil set? Which set is the best buy? How did you get your answer? What are some things other than price to consider when choosing which set to buy?

A single pencil in the eight-pencil set would cost about 25p, almost 22p in the 12-pencil set and nearly 24p in the ten-pencil set. The 12-pencil set is the best buy. Strategies may vary, but the best way may be to divide the price for each set by the number of pencils per set in order to find the cost of one pencil. Other things to consider may vary, but may include the quality and size of the coloured pencils in each set, the number of pencils you need, the amount of money you have with you.

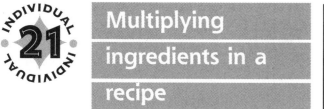

21 Multiplying ingredients in a recipe

Mental calculation strategies (× and ÷)

Molly gathered the ingredients she needed to make her famous breakfast sandwich. She got one egg, one slice of cheese, half a slice of ham, two slices of tomato, and two pieces of bread. Then her sister and brother asked her to make each of them a breakfast sandwich. How much of each ingredient does Molly need to make three breakfast sandwiches?

Molly needs three eggs, three slices of cheese, one and a half slices of ham, six tomato slices and six pieces of bread.

22 Determining quantity and cost

Problems involving 'real life' and money

The children at a day-care centre like fruit lollies for a snack. The lollies come in packs of 12. Each pack costs 89p. There are 234 children at the day-care centre. How many packs do the staff need to buy so that each child gets one lolly? How much will it cost to buy that many packs? Do the lollies cost more or less than 10p each?

234 ÷ 12 = 19r6, so the staff need to buy 20 packs to have enough for each child to get one lolly. It will cost 20 × 89p, or £17.80. The lollies cost less than 10p each: 12 × 10p is £1.20, and each package costs only 89p.

ACROSS THE CURRICULUM

HISTORY

Early cars

When cars were first invented, in the early 1900s, the top speed was about 25 miles per hour. Children can research what it must have been like on a car journey in the early 20th century. About how long would a 100-mile drive have taken in 1900? What are some things that might have slowed the motorist down in those days?

A 100-mile drive would have taken at least four hours. Things that might have slowed the motorist down may vary, but may include: tyres went flat quite often; there were no paved roads, and gravel and dirt roads often had terrible potholes; there were places that were nearly impassable, such as swollen streams and eroded banks).

Numbers divisible by two

24 INDIVIDUAL · INDIVIDUAL

Mental calculation strategies (× and ÷)
Write the numbers from 15 to 30. Divide each one by two. Copy and complete this sentence: numbers that can be divided by two with no remainder are all _____ numbers.

15 ÷ 2 = 7r1,	16 ÷ 2 = 8,
17 ÷ 2 = 8r1,	18 ÷ 2 = 9,
19 ÷ 2 = 9r1,	20 ÷ 2 = 10,
21 ÷ 2 = 10r1,	22 ÷ 2 = 11,
23 ÷ 2 = 11r1,	24 ÷ 2 = 12,
25 ÷ 2 = 12r1,	26 ÷ 2 = 13,
27 ÷ 2 = 13r1,	28 ÷ 2 = 14,
29 ÷ 2 = 14r1,	30 ÷ 2 = 15.

Numbers that can be divided by two with no remainder are all even **numbers.**

Multiplying two-digit numbers

23 PAIRS · PAIRS

Mental calculation strategies (× and ÷)
In 1893, the original Ferris wheel was built for the World's Fair in Chicago. It was huge! There were 36 cars spaced around the wheel, and each car held 50 people. Draw a picture of the Ferris wheel. What is the greatest number of people that could ride on the Ferris wheel at one time? Do modern Ferris wheels carry more or fewer people than the original Ferris wheel?

Number	Sum of digits	Divisible by three?	Check
54	5 + 4 = 9	Yes	54 ÷ 3 = 18
132	1 + 3 + 2 = 6	Yes	132 ÷ 3 = 44
516	5 + 1 + 6 = 12	Yes	516 ÷ 3 = 172

36 × 50 = 1800 people. Modern Ferris wheels carry fewer people.

Numbers divisible by three

25 INDIVIDUAL · INDIVIDUAL

Mental calculation strategies (× and ÷)
If a number is divisible by three, it can be divided by three and there will be no remainder. Here is a way to check if a number is divisible by three:

★ *add up all the digits*

★ *if the sum is a number that is divisible by three, then the original number is also divisible by three.*

Copy the chart below and complete it, writing four numbers of your own that are divisible by three.

Numbers written in the chart will vary.

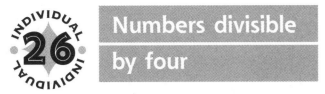

Numbers divisible by four

Mental calculation strategies (× and ÷)
Robert says that if the last two digits of a number form a number divisible by four, then the number itself is divisible by four. He used 732 as an example: 'The last two digits of 732 are three and two; 32 is divisible by four, so 732 is divisible by four: $732 \div 4 = 183$.

Write ten numbers in which the last two digits make a number divisible by four. See if Robert's rule works each time.

Numbers will vary. The rule will work each time.

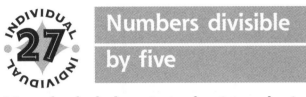

Numbers divisible by five

Mental calculation strategies (× and ÷)
The numbers 5, 10, 15 and 20 are multiples of five. Write the next ten multiples of five. What pattern do you notice?

Every number that is a multiple of five is also divisible by five. Write a rule to describe how to tell if a number is divisible by five.

25, 30, 35, 40, 45, 50, 55, 60, 65, 70; the last digit in each number is either zero or five; if the last digit of a number is zero or five, it is divisible by five.

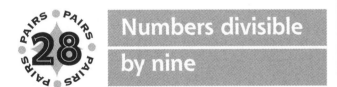

Numbers divisible by nine

Mentl calculation strategies (× and ÷)
The numbers 36, 45, 54, 63, 72, 81, 90, 99 and 108 are all divisible by nine. Write six more numbers that are divisible by nine. Look at the numbers. Write a rule for identifying when a number is divisible by nine. (Hint: you may need to use a calculator to solve this problem.) Compare your rule with someone else's rule.

Examples of numbers divisible by nine include: 117, 126, 135, 144, 153, 162. The rule should include the idea that if the sum of all the digits is nine, the number is divisible by nine.

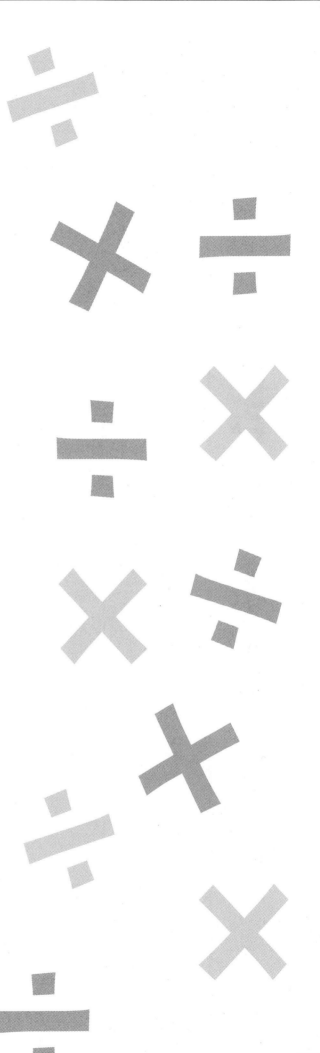

Multiplication squares

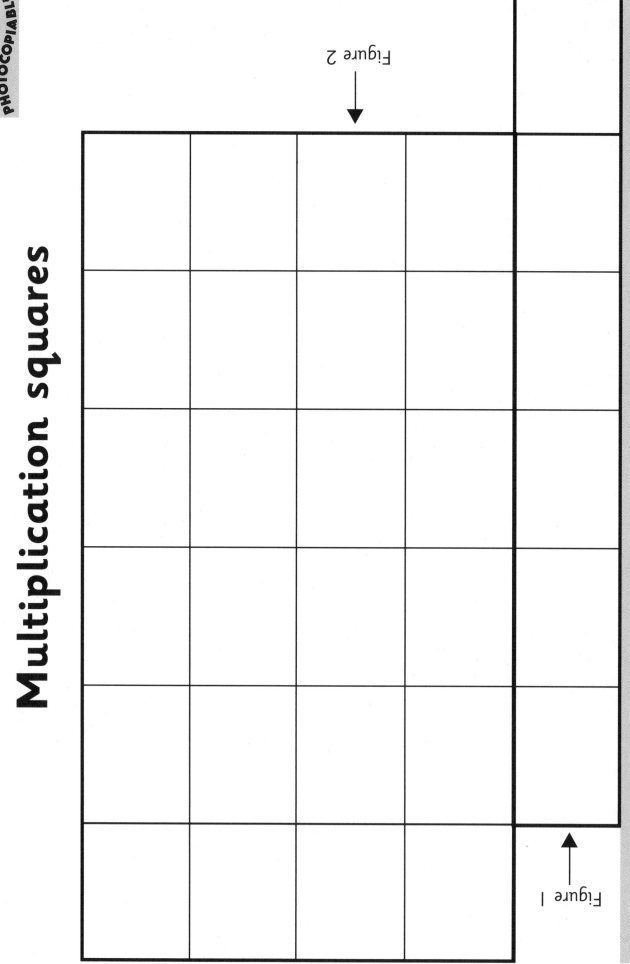

Figure 2

Figure 1

Table square

	1	2	3	4	5	6	7	8	9	10	11	12
1	1	2	3	4	5	6	7	8	9	10	11	12
2	2	4	6	8	10	12	14	16	18	20	22	24
3	3	6	9	12	15	18	21	24	27	30	33	36
4	4	8	12	16	20	24	28	32	36	40	44	48
5	5	10	15	20	25	30	35	40	45	50	55	60
6	6	12	18	24	30	36	42	48	54	60	66	72
7	7	14	21	28	35	42	49	56	63	70	77	84
8	8	16	24	32	40	48	56	64	72	80	88	96
9	9	18	27	36	45	54	63	72	81	90	99	108
10	10	20	30	40	50	60	70	80	90	100	110	120
11	11	22	33	44	55	66	77	88	99	110	121	132
12	12	24	36	48	60	72	84	96	108	120	132	144

Chapter Five

PATTERNS

1 Drawing and extending a pattern

Properties of number and number sequences

Draw nine triangles. Colour:

★ *the second and sixth triangles purple*

★ *all the odd triangles yellow*

★ *the fourth and eighth triangles red.*

If the pattern continues, what colour will the tenth triangle be? Draw it. Then continue the pattern for three more triangles.

Check the children's drawings. The colour pattern is yellow-purple-yellow-red. The tenth triangle is purple; the additional three triangles are yellow, red, yellow.

2 Identifying and extending patterns

Properties of number and number sequences

Describe what function is forming these patterns and write the next three numbers:

37, 35, 33, 32, __?, __?, __?

35, 38, 41, 44, 47 __?, __?, __?

Combine the two patterns above to create a new pattern, following these steps:

begin by writing the number 35

use the first pattern to write the next number

use the second pattern in to write the next number

repeat, alternating between first and second patterns

what are the first five numbers in the new pattern?

The first pattern is 'subtract two'; the next three numbers are 29, 27, 25. The second pattern is 'add 3'; the next three numbers are 50, 53, 56. The first five numbers of the new pattern are 35, 33, 36, 34, 37.

Properties of number and number sequences

Create a number pattern of your own. It can be an addition or subtraction pattern, an odd or even number pattern, a pattern of numbers that have common digits, or any other pattern you can think of. Swap patterns with your partner. Describe the pattern and write the next five numbers.

Patterns and numbers will vary.

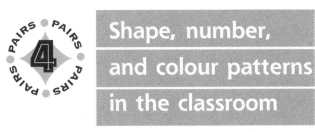

Shape, number, and colour patterns in the classroom

Shape and space

There are probably many shape, number and colour patterns in your classroom. For example, the ceiling tiles could have patterns of dots – this is a shape pattern. The calendar may show the numbers 7, 14, 21, 28 in one column – a number pattern. The noticeboard could have a blue and white border around it – a colour pattern.

Work with your partner to find a shape, a number, and a colour pattern in your classroom. Draw a picture or write a description of each one.

Patterns and descriptions will vary.

Shape and space

Look at the shape pattern in figure 1 on photocopiable page 49 for 30 seconds. Turn the page over so you can't see the shapes, and describe the pattern to yourself. Look at the pattern again for ten seconds, then turn the paper over again. From memory, draw the next three shapes in the pattern. Compare your drawing to figure 1.

Children's drawings should show three shapes the same as the first three shapes in figure 1 on photocopiable page 49.

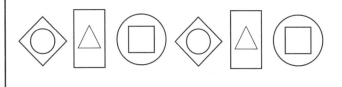

Extending patterns from memory

Shape and space

Look at the shape pattern in figure 2 on photocopiable page 49 for 30 seconds. Turn the page over so you can't see the shapes. Close your eyes and try to picture the pattern in your mind. Look at the pattern again for ten seconds, then turn the paper over again. From memory, draw the next row in the pattern. Compare your drawing to figure 2 on photocopiable page 49.

Children's drawings should show a row of shapes identical to each row in figure 2 on photocopiable page 49.

ACROSS THE CURRICULUM

LITERACY & MUSIC

Patterns in poetry

Poems often have sound patterns created by rhyme and rhythm. Write a simple poem such as the following on the board:

Jack and Jill went up a hill
to fetch a pail of water
Jack fell down and broke his crown
and Jill came tumbling after.

Ask children to read the poem aloud and to identify the pairs of rhyming words (Jill, hill; down, crown; water, after). They should then read the poem again, clapping softly to identify the rhythm in each line.

Working in pairs, ask children to write a short poem, then to read their poems aloud. Ask the other children to identify the rhyme and rhythm patterns.

Day of the week	Mon	Tue	Wed	Thu	Fri
Number of tents	2	4	6	8	10

Drawing and extending a pattern

Shape and space

Cut out the shapes in figures 1 and 2 on photocopiable page 49. Take turns with your partner to mix up the shapes, arranging them to create a new pattern. Get your partner to describe the pattern you have created. Draw each pattern you make.

Patterns shown in drawings will vary.

Finding a pattern to solve a problem

Properties of number and number sequences

Nasser and Judy went on a camping trip. On Monday they saw some tents near their campsite. On Tuesday they saw twice as many tents as the day before, and on Wednesday three times as many tents as on Monday. On Thursday they saw eight tents, two more than on Wednesday. On Friday they saw two more tents than on Thursday, which was five

times as many as on Monday.
Make a table. Describe the pattern you see. How many tents did they see on Monday? If the pattern continues, how many tents will they see on Saturday?

The number of tents increases by two each day; they saw two tents on Monday; they will see 12 tents on Saturday.

Fibonacci sequence

Properties of number and number sequences

There is a special pattern of numbers called the Fibonacci sequence, named after the man who discovered it. He studied the natural world and noticed a pattern in the numbers of things, such as petals on flowers and leaves on stems. This is how the Fibonacci sequence begins: 1, 1, 2, 3, 5, 8, 13, 21, 34.

Describe the pattern. Discuss it with your partner. Predict what the next number in the sequence is.

Each number in the sequence is the sum of the two numbers before it. Predictions will vary; the next number is 55.

ACROSS THE CURRICULUM

Sound and movement patterns

Rhythm is very important to dancers and musicians. Rhythm is conveyed in the pattern of sounds in music and the pattern of movement in dance. Get children to work together in small groups to create a rhythm using sound (by tapping, clapping, or stamping), or a rhythm using movement. Each group should demonstrate their sound or movement pattern and teach it to the rest of the class.

10 TEACHER LED

Finding examples of the Fibonacci numbers

Properties of number and number sequences

Take the children for a nature walk around the school grounds and look for things that represent one or more numbers of the Fibonacci sequence – 1, 1, 2, 3, 5, 8, 13, 21, 34, 55, 89 and so on. For example, the number of petals on a flower may be five or eight, or the leaves on a stem might be clustered in groups of five. Get the children to record what they find on the walk. Back in the classroom, make a class list of the items found and the Fibonacci numbers they represent.

Answers will vary.

11 INDIVIDUAL

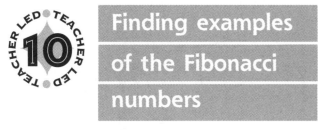

Illustrating numbers from the Fibonacci sequence

Properties of number and number sequences

Draw a picture of something from nature that represents numbers from the Fibonacci sequence. For example, a stem with two leaves on the left side, then three leaves on the right, then two leaves on the left, and one on the right. On the stem is a flower with 13 petals.

Answers will vary.

12 PAIRS

Pictograms

Shape and space

Native Americans drew pictograms to tell stories, describe events, or record information. Pictograms used symbols for animals, people, and objects. Without using a symbol for the number four, how could you represent four deer in a pictogram? Write a sentence about yourself or your family that includes a number, for example: 'Our family has three guinea pigs'. Draw a pictogram to represent your sentence. Swap papers with your partner and write the sentence shown by his or her pictogram.

Draw four of the symbols used to represent a deer. Sentences and pictograms will vary.

Musical patterns (Music)

Explain that manufactured objects also often exhibit Fibonacci proportions. Use piano keys as an example (although there are 88 instead of 89 keys): there are clusters of two black keys above three white keys and three black keys above five white keys; octaves are made up of eight keys.

Invite volunteer musicians to demonstrate and discuss patterns in music with students.

13 Using a number–letter code 1

Properties of number and number sequences

Follow these simple directions to turn words into number codes using the code wheel on photocopiable page 50:

Cut the wheels out, and make a hole through each centre. Place the smaller wheel on top of the larger one, lining up the holes. Put a brass paper fastener through the holes to keep the wheels together.

Turn your code wheel so the number eight lines up with the letter A. To write the number code for the word 'hat', write the numbers that line up with the letters H, A, and T. What is the number code for 'hat'?

Keep the number eight lined up with the letter A on your code wheel. Write a sentence, using the number code for each word in the sentence. Swap coded sentences with your partner. Decode your partner's sentence.

The number code for hat is 15-8-1; sentences will vary.

14 Using a number–letter code 2

Properties of number and number sequences

Follow the directions as the activity above to construct a code wheel (see photocopiable page 50), then decode this sentence. (Hint: 23, 7 is the number code for the word 'is'.)

8, 3, 18, 15, 13   23, 7   13, 3, 9, 6   26, 9,

17, 25, 13   18, 15, 13

The sentence is: *today is your lucky day.*

15 Creating a code

Properties of number and number sequences

Assemble your code wheels (see the directions in activity 13) to make up your own code. You can write numbers for letters, or you can use other symbols. Write a sentence using your code. Then write a key that describes how to decode it. Swap papers with your partner. Decode the sentence he or she wrote.

Codes and sentences will vary.

Patterns

Figure 1

Figure 2

Code wheel

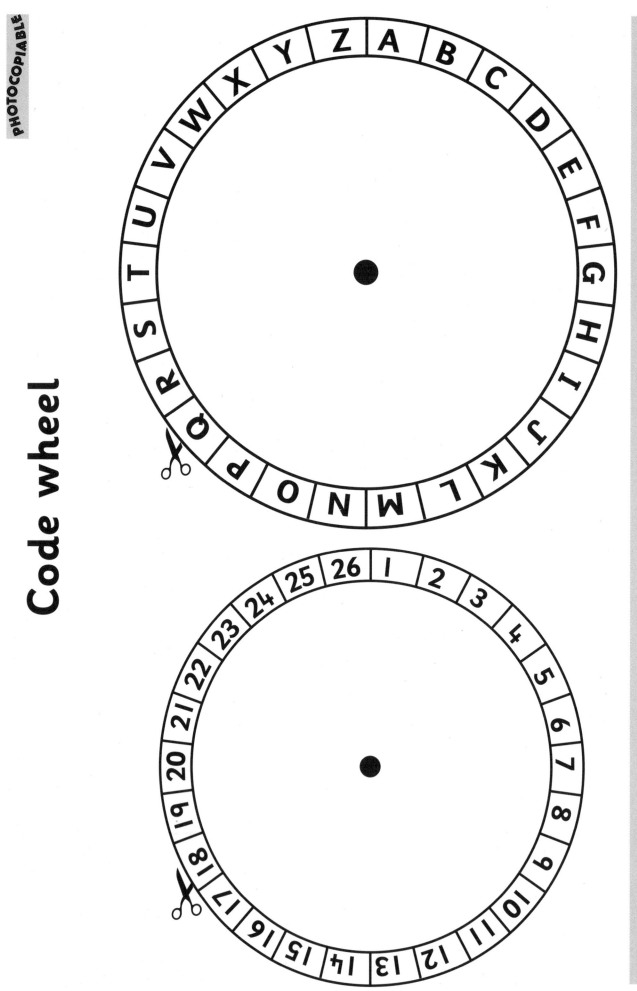

Chapter Six

FRACTIONS, DECIMALS and PERCENTAGES

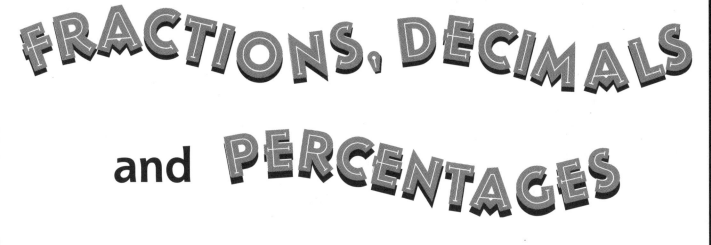

1 INDIVIDUAL

Paper-folding: halves

Fractions

Fold a sheet of paper in half. How many sections do you have when you unfold it? How many sections do you think you will have if you fold the paper in half, then in half again? And in half a third time? Fold the paper to check your predictions. Will the number of sections be the same with paper of any size and shape?

Two sections; predictions will vary. A paper folded in half twice will have four sections. A paper folded in half three times will have eight sections; yes, the number of sections would be the same for any size and shape of paper.

2 INDIVIDUAL

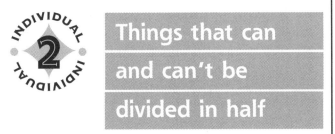

Things that can and can't be divided in half

Fractions

List three things that would be just as good if you cut them in half. List three things that would be ruined if you cut them in half.

Answers will vary, but may include: biscuits, sandwiches, oranges; for the second question, answers will vary but may include: money, clothing and books.

3 INDIVIDUAL

Applying the concept of one third

Fractions

Draw a circle and divide it into three equal sections. Shade one section. One third is the name of the fraction that describes the shaded part. Tell a story that has the fraction one third in it.

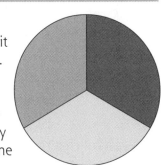

Check children's circles; stories will vary.

Fractions equal to one

Fractions

Cut out each fraction strip from photocopiable page 57. Then cut up each strip into equal sections, cutting along the dotted lines. You should have two 'one half' sections, three 'one third' sections, four 'one quarter' sections and so on.

Cover up the strip representing one whole (labelled 1) with three different combinations of fraction sections. Record the combination by writing number sentences, for example: one half + one quarter + one eighth + one eighth = one.

Number sentences will vary, but may include: one half + one third + one sixth; one third + one sixth + one sixth + one ninth + one ninth + one ninth; one half + one quarter + one twelfth + one twelfth + one twelfth.

Comparing fractions

Fractions

Cut out each of the fraction strips from photocopiable page 57 to play the following fraction comparison game. One partner picks two sections. The other partner compares the sections and records the comparison using >, <, or =. Continue taking turns, choosing different fractions and comparing them in writing until both pairs of children have managed to record five comparisons.

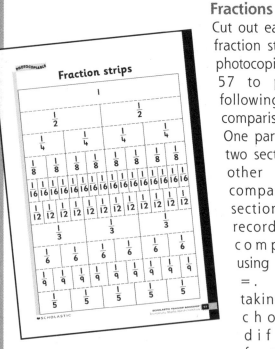

Recorded comparisons will vary.

Finding equivalent fractions

Fractions

Cut out each fraction strip from photocopiable page 57. Follow these steps to find equivalent fractions:

1   *One partner chooses a fractions section.*

2   *The other partner tries to find two or more fraction sections that, when placed together, are equivalent, or the same size as, the first fraction section.*

3   *If the partner finds two or more sections that are equivalent, he or she records it in a number sentence. For example: one eighth + one sixteenth + one sixteenth = one quarter.*

4   *If the partner cannot find two or more sections that are equivalent, he or she records the fraction and writes: No equivalent sections found.*

5   *Continue taking turns choosing and finding equivalent sections until each partner has written five number sentences.*

Number sentences will vary. Children will not be able to find two or more sections equivalent to one fifth, one ninth, one twelfth or one sixteenth on the photocopiable sheet.

Fraction of a number

Fractions

Chris and his dad are making spaghetti for dinner. The recipe makes enough spaghetti for 12 people, but they only need to make enough for one quarter of that number. For how many people are they making spaghetti?

These are the ingredients from the recipe:

1600ml tomato sauce

800ml tomato purée

2400ml chopped tomatoes

2000g spaghetti

Rewrite the quantities of ingredients so Chris and his dad will make just the amount of spaghetti they need.

Three people; 400ml tomato sauce, 200ml tomato purée, 600ml chopped tomatoes, 500g spaghetti.

Adding fractions

Fractions

Kiley has a little dog, Shadow, and Sophie has a very big dog, Rollo. Shadow eats one third as much as Rollo. Rollo eats one bag of dog food each week. How much does Shadow eat each week? How much

does Rollo eat in eight weeks? How much does Shadow eat in eight weeks? Work with your partner to make a table to find the answers.

Shadow eats one third of a bag of dog food in a week; Rollo will eat eight bags of food in eight weeks; Shadow will eat two and two-thirds bags of food in eight weeks. One possible table:

Number of bags of dog food eaten

Week	Rollo	Shadow
1	One	One-third
2	Two	Two-thirds
3	Three	One
4	Four	One and a third
5	Five	One and two-thirds
6	Six	Two
7	Seven	Two and a third
8	Eight	Two and two-thirds

ACROSS THE CURRICULUM

PSHE & SCIENCE

Fraction feast

Ask children to bring in food items that can be cut into many equal pieces, such as bananas and apples, brownies (or anything else baked in a rectangular pan), cheese and pizza. Discuss the variety of the shapes and sizes of the food. Help the children to cut the foods up; as each item is cut into equal pieces, discuss the fractions represented. For example, if an apple is cut into six wedges, one wedge represents one sixth of the apple, two wedges one third of the apple and so on.

9 INDIVIDUAL

Dividing a lesser number by a greater number

Pencil and paper procedures (× and ÷)

Suppose you have three friends over to play, and everybody (including you) wants a snack but there are only three cupcakes. Draw a picture to show how you will divide the cupcakes so each person gets an equal share.

Drawings may vary, but should show that each person would receive three quarters of a cupcake.

10 PAIRS

Comparing fractions

Fractions

Use what you know about fractions to write these fractions in order from least to greatest: four fifths, three quarters, seven eighths, two thirds. Compare your answers with your partner's.

Describe one way you could check your answers.

Two thirds, three quarters, four fifths, seven eighths; ways to check answers will vary, but may include: comparing the fractions using the strips on photocopiable page 57; writing an equivalent fraction for each using the lowest common denominator (12) and comparing the numerators.

11 INDIVIDUAL

Fractions equal to one

Fractions

John, Jamie and Joel are brothers who worked together to paint a fence. John is the oldest and said he'd paint half the fence. Jamie is the middle brother and said he'd paint one third of the fence. Joel, the youngest, would paint whatever part was left. Draw a picture to show how much of the fence each brother painted. Then write a number sentence describing your picture.

Drawings should show that John painted half of the fence, Jamie one third and Joel one sixth; one half + one third + one sixth = one.

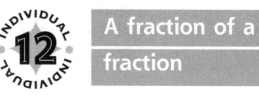

12 INDIVIDUAL

A fraction of a fraction

Fractions

Draw a picture to show that one quarter of one half is the same as one half of one quarter.

Pictures may vary. One possible picture:

13 INDIVIDUAL

Finding half of a fraction

Fractions

Derek and Tony decide to combine their money to buy a pizza. Tony has only half as much money as Derek, so he is only going to eat half as much pizza. How much of the pizza will each one eat? Draw a picture to find the answer.

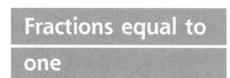

Derek will eat two thirds of the pizza and Tony will eat one third; pictures will vary.

Ratios

Ratios

Make a design showing diamonds and triangles in a 'one to four ratio' – that is for every diamond you draw, draw four triangles. If you draw two diamonds, how many triangles should you draw?

Make another design with diamonds and triangles, but use a different ratio. Swap drawings with your partner and write the ratio it shows.

Eight triangles; designs will vary; ratios will vary.

ACROSS THE CURRICULUM

Enlarging a drawing

Using a copy of photocopiable page 58, ask children to enlarge the drawing of the make-believe creature using a 1:3 ratio. Explain that each part of the monster drawn on the small grid will be drawn 3 times larger on the big grid. Instruct them to draw what is in each square of the small grid of the corresponding square in the big grid, beginning with the square in the top left-hand corner.

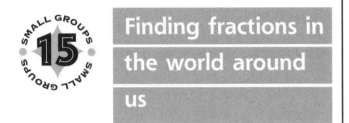

Finding fractions in the world around us

Fractions

Ask groups of children to brainstorm the ways that fractions are used inside and outside the classroom. Get groups to share their findings and record them on a class list.

Answers will vary, but may include: telling time – half an hour, quarter of an hour; measuring – quarter kilometre, half litre; playing music – half notes, eighth notes.

Writing decimals

Decimals

Tina used a spinner to get these digits: 5, 9, 2, 7, 4, 1. She is going to use the digits to write decimals. The only rule is she must have at least one digit on either side of the decimal point.

What is the largest possible decimal Tina can write using all the digits? What is the smallest possible decimal Tina can write using all the digits?

97542.1; 1.24579.

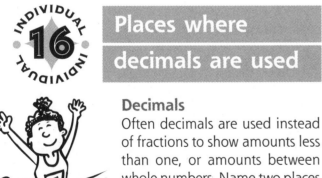

Places where decimals are used

Decimals

Often decimals are used instead of fractions to show amounts less than one, or amounts between whole numbers. Name two places where decimals are used, and give three examples of the decimals that may be used there.

Answers will vary, but may include: at the deli counter in a supermarket – 0.75kg, 1.25kg; at athletic events – a long-jump of 7.54m; buying fabric to make clothes – 3.5m.

Finding percentages

Percentages

If 20% of the people in a city drive red cars, what percentage of people in the city do not drive red cars?

Make up a question of your own that has an answer of 60%.

80% of the people do not drive red cars; questions may vary, but should include a combination of numbers that add up to 100%.

19 Drawing to show percentages

Percentages

Circle the four squares made up of 100 smaller squares on the 'Fractions and decimals' worksheet (page 59). Label each of these squares a, b, c, or d, and in each square draw one of the following:

★ a design, 30% of which is coloured blue

★ the letter 'L' so that it covers 14% of the square

★ a design, of which 25% is yellow and 75% is green

★ a house that covers 60% of the square.

Check children's drawings.

20 Drawing to show equivalent fractions and decimals

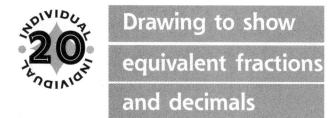

Fractions and decimals

Use the 'Fractions and decimals' worksheet (page 59). In the two squares in Section 1, colour in one quarter of the top square and 0.25 of the bottom square. What do you notice?

Colour in the same amount in each pair of squares in the next three sections to show other equivalent fraction and decimal pairs. Write each equivalent fraction and decimal pair shown.

The same amount of each square is coloured; equivalent fraction and decimal pairs will vary.

21 Writing percentages and fractions as decimals

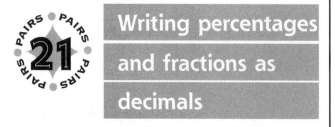

Fractions, decimals and percentages

There is a lot of water in the food we eat. A watermelon is 98% water. A potato is three-quarters water. An apple is four-fifths water. Write these foods in order from the least amount of water to the greatest amount. (Remember, three quarters means the same as $3 \div 4$.) Compare your answer with your partner's. (If necessary, you may use a calculator to solve this problem.)

Potato, apple, watermelon.

22 Adding and multiplying decimals

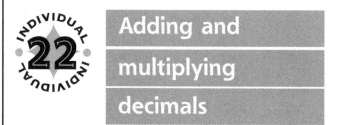

Decimals

Bobby is earning money for charity by doing a walk-a-thon. For every kilometre he walks, he earns 75p for the charity. In the first hour he walked 5.2km, then 5.7km in the second hour, 6.1km in the third hour and 6.2km in the fourth. How many kilometres did he walk? How much money did he earn for the charity? (If necessary, you may use a calculator to solve this problem.)

21.2km; £15.90.

23 Writing percentage problems

Percentages

Make up a quiz on percentages to give to your partner. Pose questions such as: 'Five is 50% of what number?'

The answer to the question is ten. Make up five questions in all. Swap quizzes, then later correct the quizzes together.

Questions will vary.

Fraction strips

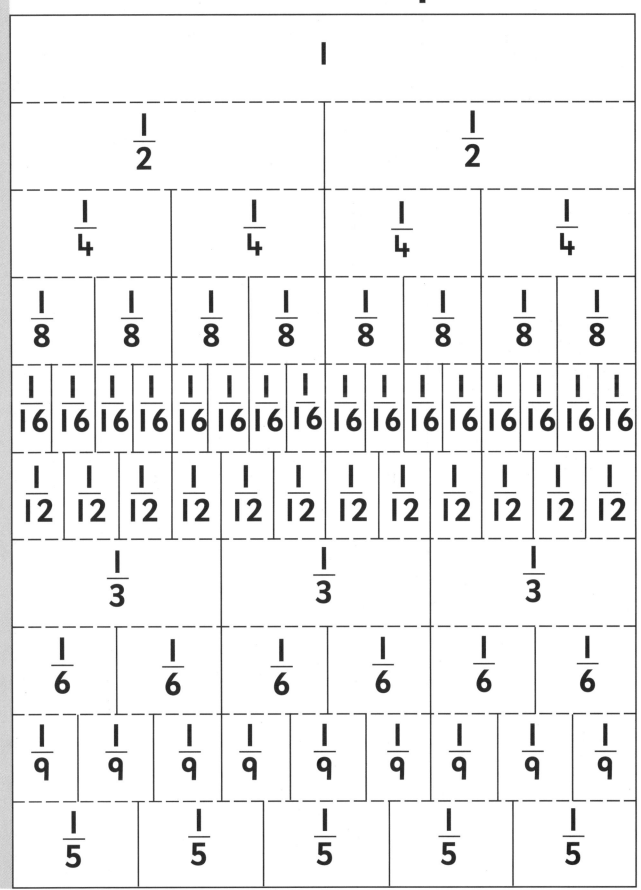

Enlarging a drawing

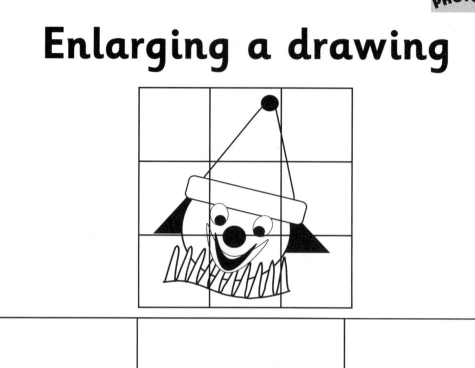

Fractions and decimals

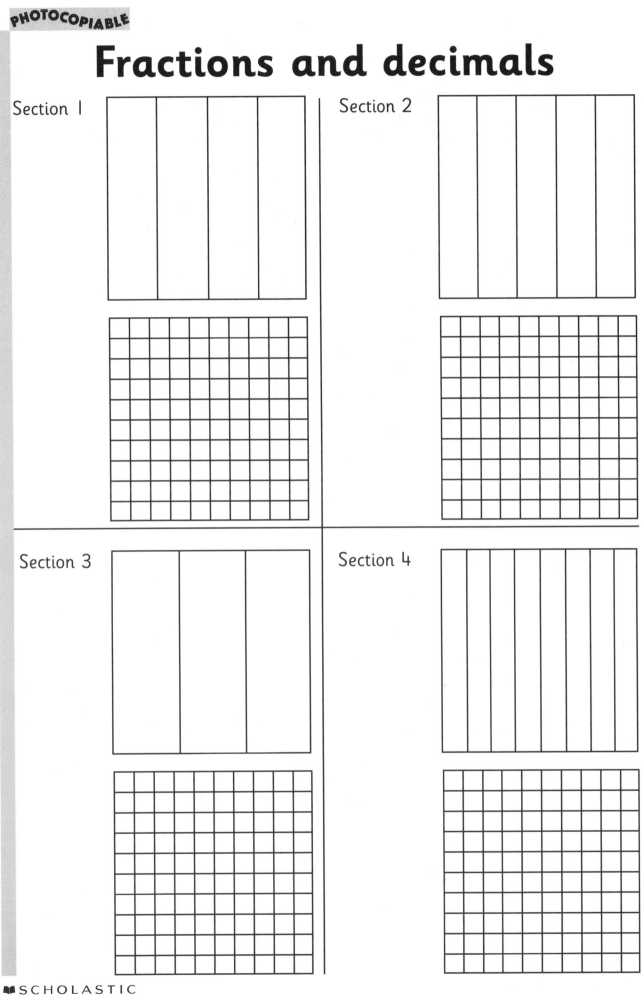

Section 1

Section 2

Section 3

Section 4

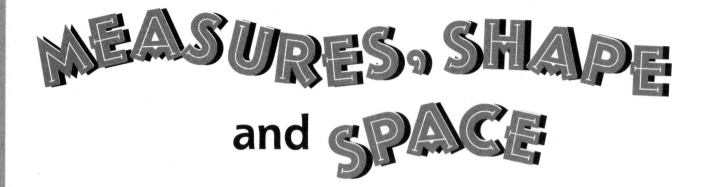

Chapter Seven

MEASURES, SHAPE and SPACE

1 Identifying shapes and estimating sizes

Shape and space
Draw and label as many things as you can that are the same shape as a ten pence coin. Estimate whether each one is smaller or larger than the coin. Compare the items to a real coin to check your estimates.

Items and estimates will vary, but may include: the face of a watch or clock, the bottom of a pencil or marker, or the bottom of a drinking glass.

2 Estimating size

Shape and space, measures
Draw the following items from memory (no looking!), trying to make them actual size: paper clip, pen, ruler. Compare your drawings to the real items. How did you do in estimating the sizes?

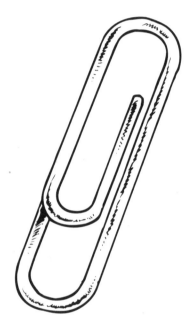

Comparisons will vary. Children should indicate whether the drawn objects were larger, smaller, or the same size as the actual items.

3 Identifying rectangles and their attributes

Shape and space
Make a list of everything you see, from where you are sitting, that has a rectangular shape. Describe two ways in which all rectangles are the same.

Lists will vary. Descriptions of rectangles will vary but should include two of the following: four sides, two pairs of parallel sides, four right angles or square corners.

Identifying a shape from its description

Shape and space

Look at the four-sided shapes in figure 1 on photocopiable page 67.

1   Write the number of the shape that fits the following description: each of its corners is a square corner, or right angle. Two of the sides are twice as long as the other two sides.

2   Draw one of the other four-sided shapes from figure 1 and write a description.

---

1 Number 3. 2 Drawings and descriptions will vary, but may include: number 1 – each of its corners is a square corner, or right angle, and all four sides are the same length; number 2 – its corners are not all the same size, all four sides are the same length; number 4 – its corners are not all the same size, only two of the sides are the same length.

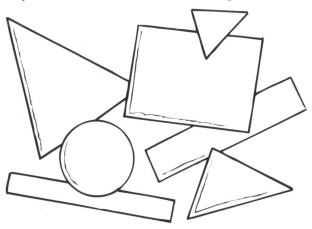

Describing and identifying shapes

Shape and space

Take turns describing a shape in figure 1 (page 67) for your partner to identify. Without mentioning the name of the shape, you can describe the sides as slanted or straight, the lengths of sides, or sizes of angles, and so on. Give only one clue at a time. See how many different ways you can describe each shape, and how few clues are needed to identify it.

Answers will vary.

Drawing triangles

Shape and space

Draw each triangle described below. Label each one with the same number as its description – 1, 2 or 3:

1   none of the three sides are the same length;

2   two of the three sides are the same length;

3   all three angles or corners are the same size.

Drawings should show: 1 a scalene triangle; 2 an isosceles triangle; 3 an equilateral triangle.

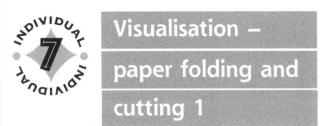

Visualisation – paper folding and cutting 1

Shape and space

Suppose you folded a piece of paper in half and made three cuts along the fold, as shown in figure 2 on photocopiable page 67. What shapes do you think you would see when you unfolded the paper? Draw your prediction.

Now fold and cut a piece of paper as shown. What shapes do you see when you unfold the paper? How do the results compare with your predictions?

**Predictions will vary.
The top shape is a triangle, the middle shape is a rhombus, and the bottom shape is an oval.**

Visualisation – paper folding and cutting 2

Shape and space

Take a piece of paper and fold it in half. Make some cuts along the fold. Predict, in writing or by drawing, what the paper will look like when you unfold it. Unfold the paper to check your prediction.

Repeat the activity, this time folding a piece of paper in half twice.

Write a few sentences describing how the unfolded pieces of paper compared with your predictions.

Predictions and sentences will vary.

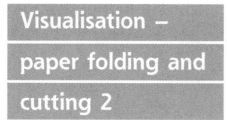

ACROSS THE CURRICULUM

LITERACY

Shape poetry

Ask children to write poems describing objects. Model how to arrange the words of the poem into the shape of the object using the example given below. Display the poems for everyone to enjoy.

> All
> diamonds shimmer
> their light is reflected into
> a thousand points of
> colour

Paper folding – sixths

Shape and space

Divide a sheet of paper into six equal sections by drawing five lines. Divide another sheet of paper into six equal sections by drawing only three lines.

Answers may vary.

Paper folding – cube

Shape and space

Look at the shape in figure 1 on the silhouette page (photocopiable page 68). Think about how you could fold it into a cube. Test your idea by cutting out the shape and folding and taping it into a cube. Write or draw the steps you took to do it.

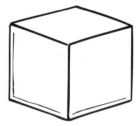

Answers and drawings will vary.

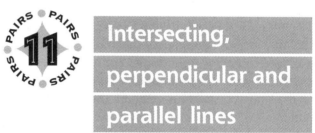

Intersecting, perpendicular and parallel lines

Shape and space

Two lines on a plane – a two-dimensional surface – can intersect or be parallel to one another. Some lines that intersect may be perpendicular to each other.

Use a ruler, a sheet of paper, and a red and green marker. Take turns drawing lines with your partner. Draw the lines going in different directions, from one edge of the paper to another edge.

Take turns drawing circles around points where lines intersect, outlining perpendicular lines in red, and outlining parallel lines in green. Continue until you have found all the ways the lines you drew relate to one another.

Check the children's drawings: intersections (circled) are the points where lines meet; perpendicular lines (red) intersect at right angles; parallel lines (green) do not intersect – the lines are always the same distance apart.

Missing angles

12 INDIVIDUAL

Shape and space

Look at the pie and the pie pieces labelled A, B, and C in figure 2 on the silhouette page (photocopiable page 68). Think of a way, other than cutting out the pie pieces and fitting them into the empty space, to tell which piece is the one missing from the pie. Try your idea. Which piece is it?

Answers may vary. Possible answers include: measure the angle of the empty space in the pie and find the piece that has that angle; or trace the outline of the empty space in the pie and match its outline to one of the pieces. Piece A is the missing piece.

Properties of polygons

13 INDIVIDUAL

Shape and space

Polygon means 'many angles' in Greek. Polygons are closed figures that have three or more angles and three or more sides. Draw three different polygons. Identify the number of sides and angles in each one. What do you notice?

Polygons drawn will vary: the number of sides in each polygon is equal to the number of angles.

Names of polygons

14 INDIVIDUAL

Shape and space

Write the meaning of each of the following prefixes. You can use a dictionary to help you.

penta- hexa- hepta- octa- nona- deca-

Look at the shapes on the 'Polygon card' worksheet (photocopiable page 69). Write the name of each polygon on it. Choose from the following names: *pentagon, hexagon, heptagon, octagon, nonagon,* and *decagon.*

Penta – 5; hexa – 6; hepta – 7; octa – 8; nona – 9; deca – 10; the pentagon is the five-sided shape, the hexagon the six-sided shape and so on.

Polygons race

15 TEACHER LED

Shape and space

Divide children into teams of four or five. Explain that each team should name as many examples of polygons as they can in five minutes. The team must record the name 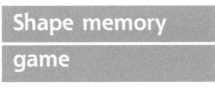 of the polygon and an example or examples. For instance: triangle – a 'Give way' sign.

At the end of five minutes, teams should share their lists. If a team lists an example that one or more of the other teams has, they score five points. If a team lists an example that no other team has, they score ten points. The team with the most points wins.

Examples will vary, but may include: rectangles – doors and windows, pentagons – arrow road signs.

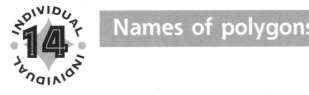

Shape memory game

16 PAIRS

Shape and space

With a partner, cut out the cards from the 'Polygon card' worksheet (photocopiable page 69). Player one chooses some of the shapes and displays them to his or her partner for an agreed amount of time, such as ten seconds. Then cover the shapes. Player two tries to name the cards that were shown and identify the ones that were missing. Uncover the cards to check the answers. Swap roles and play again.

Answers will vary.

SCHOLASTIC TEACHER BOOKSHOP

Concentric shapes

Shape and space

Concentric shapes have a common centre. To draw concentric shapes, outline a large shape on a piece of blank paper. (You can draw a shape with a combination of curved and straight lines, like a square with a half circle above it, or any shape you'd like.) Draw the same outline, only smaller, inside the first outline. Continue drawing smaller and smaller outlines of the shape until you have no more room to draw. How many concentric shapes did you draw?

Check children's drawings; the number of concentric shapes drawn will vary.

Concentric shapes in nature

Shape and space

Name something in nature that has concentric shapes. Draw it.

Answers and drawings will vary, but may include: a spider's web, ripples made in a pond when a stone is thrown into it, age rings in a log.

Tessellations

Shape and space

Tessellations are like puzzles. They are patterns made with shapes that fit together without overlapping or leaving spaces. When shapes fit together this way, we say they tessellate.

Cut out the pentagon, hexagon, and octagon shapes from the 'Polygon card' worksheet (page 69). Using the pentagon, trace it repeatedly, and see if you can draw the outlines so they tessellate, or fit together like pieces of a puzzle. Try it with the heptagon and octagon. Which shapes tessellate?

The pentagon and hexagon can be made to tessellate.

Composite shapes

Shape and space

You can create shapes out of other, smaller shapes. For example, it is possible to arrange six triangles to make a hexagon.

Use pattern blocks, or draw some shapes of your own. Draw and label the combinations of smaller shapes that make up larger shapes you find. (If you're using pattern blocks, trace them for your drawing). For example, show a square with a line drawn across to divide it into two triangles. Write: *Two triangles make one square.*

Answers will vary.

Investigate the artists

Abstract artists use line, shapes and tessellations in interesting ways. Ask small groups to research and report on artists such as Pablo Picasso, Piet Mondrian, Georges Braque, Henri Matisse and Alexander Calder.

Plan a field trip to an art gallery so children can see examples of the connection between maths and art. In addition to abstract drawings and paintings, objects such as tapestries and vases often have patterns showing tessellations and concentric shapes.

Geometric designs

Shape and space
Draw a picture or design that includes concentric shapes and tessellations. Show your design to your partner and ask him or her to find an example of each type of pattern.

Drawings will vary.

An introduction to tangrams

Shape and space
Tangrams were invented by the Chinese 4000 years ago. Legend suggests that a man named Tan was carrying a ceramic tile for the emperor, when he dropped it. He found many wonderful figures and designs while he was trying to reassemble the tile. The emperor didn't get his tile, but Tan became the famous inventor of tangrams. All tangrams are made up of seven pieces that can fit together to form a square.

Look closely at figure 1, the tangram puzzle, on the 'Tangrams' worksheet (photocopiable page 70),

to see how the pieces fit together. Cut out the pieces and then put the square back together. Outline each tangram piece in the square you made to show where each piece belongs.

See the 'Tangrams' worksheet for the way the pieces should fit together.

Describing tangram pieces

Shape and space
Make a list of the shapes of the tangram pieces from photocopiable page 70.

1   How many different shapes are there?

2   How many different sizes of triangles are there?

3   What pieces, when placed together, form a triangle that matches a large triangle?

---

1 There are three shapes – triangle, square, and rhombus; 2 there are three different sizes of triangles; 3 two small triangles and the square form a triangle congruent to a large triangle.

D&T

Origami (D&T)

Origami is the Japanese art of paper-folding. Many beautiful three-dimensional objects can be made by folding paper. In Japan, very young children learn the skill of origami.

To make your own origami cup, cut out the large square from the 'Origami cup' worksheet (photocopiable page 71). Copy the letters A–F as they appear on the front of the square onto the back of the square too, so that when the paper is folded you can read the letter no matter which side is showing. Follow the steps shown to make a cup. (If you'd like to be able to drink out of the cup, make it out of waxed paper.)

D&T

Paper flowers (D&T)

Brighten up the classroom with some colourful tissue- or crêpe paper flowers. Follow the directions to make one flower:

1 Make a stack of four or five square pieces of tissue or crêpe paper.
2 Fold the stack of papers accordion-style.
3 Twist a pipe-cleaner or piece of florist's wire around the middle of the paper, making a bow tie.
4 Fluff out the layers.

24 Making tangram designs

Shape and space

Make a design with your tangram pieces. Show it to your partner and ask them to copy it using his or her own tangram pieces. Take turns making and copying designs.

Designs will vary.

25 A tangram fox

Shape and space

Look at the fox in figure 2 on the 'Tangrams' worksheet (photocopiable page 70). Use your seven tangram pieces to make this animal. Outline each tangram piece in your fox shape to show where each piece belongs.

26 A tangram dog

Shape and space

Change the fox you made in the previous activity into the dog in figure 3 on photocopiable page 70 by swapping two pieces and turning all the other pieces in place. Outline each tangram piece in your dog shape to show where each piece belongs. Colour the two pieces you swapped yellow. Colour the pieces you turned in place green.

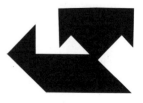

Shapes

Figure 1

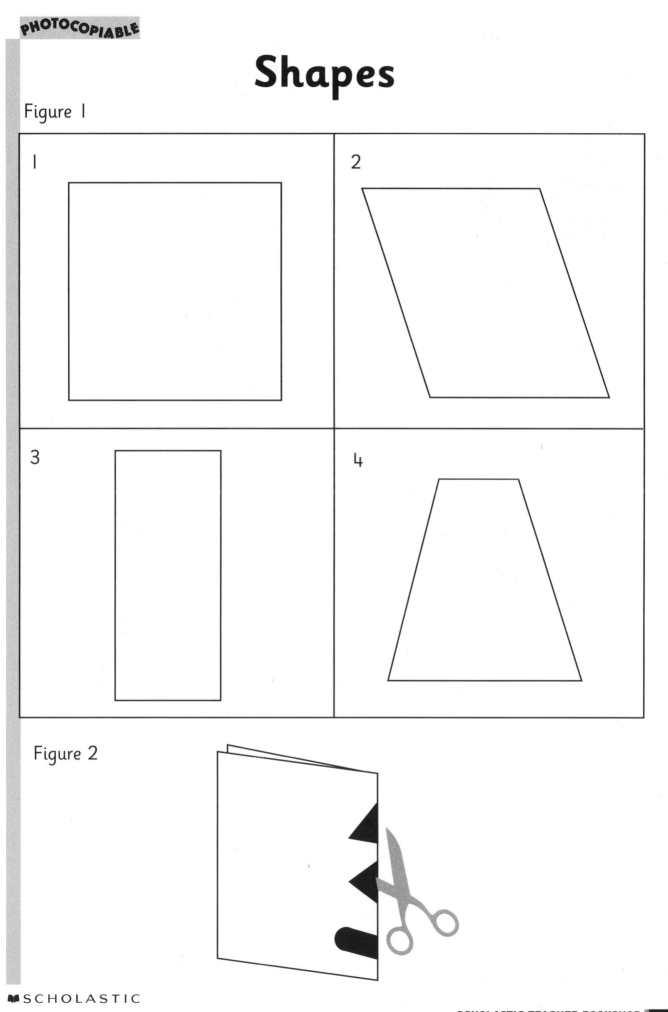

Figure 2

Silhouettes

Figure 1
Cube

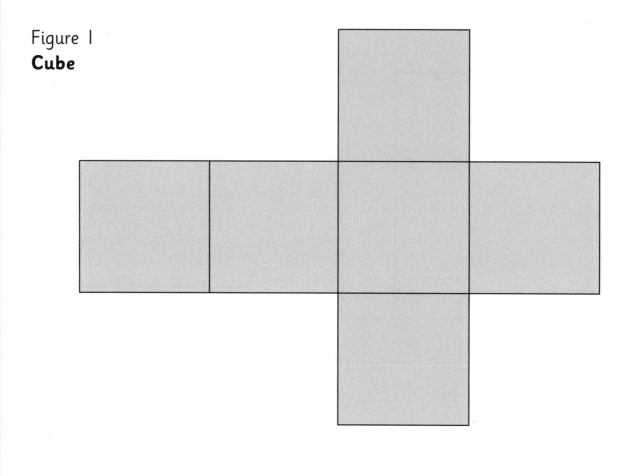

Figure 2

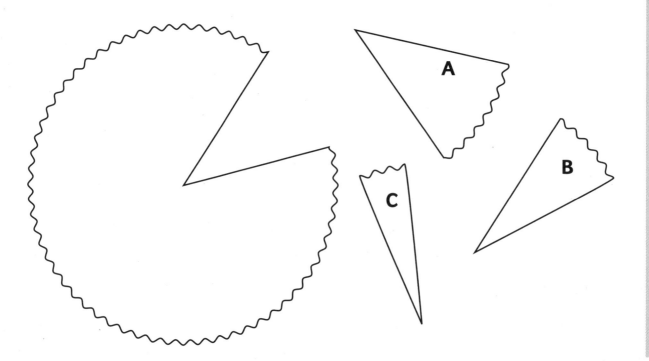

📖SCHOLASTIC

Polygon card

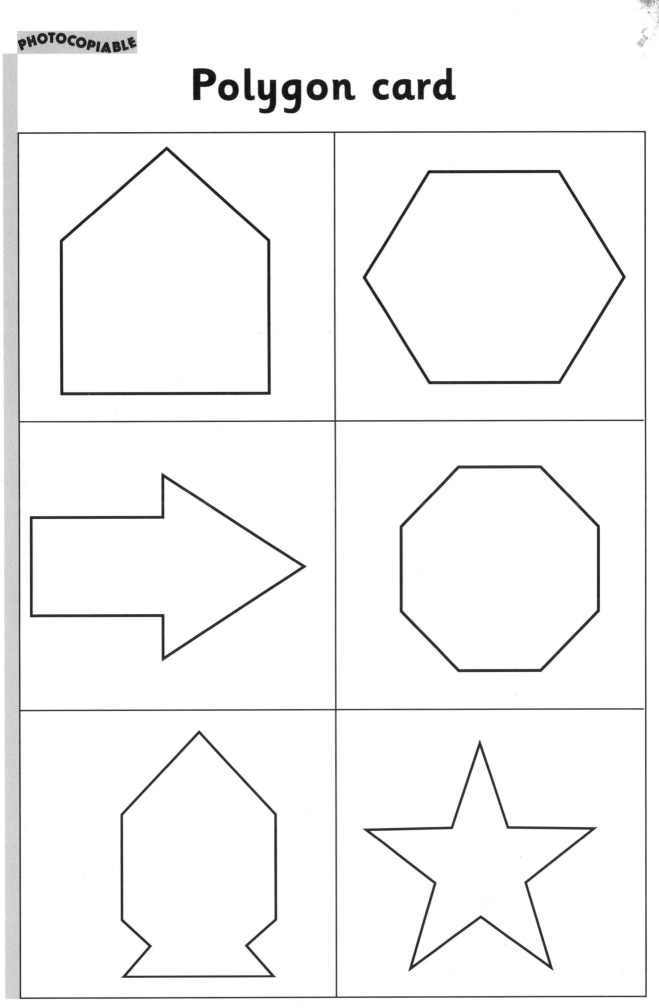

Tangrams

Figure 1

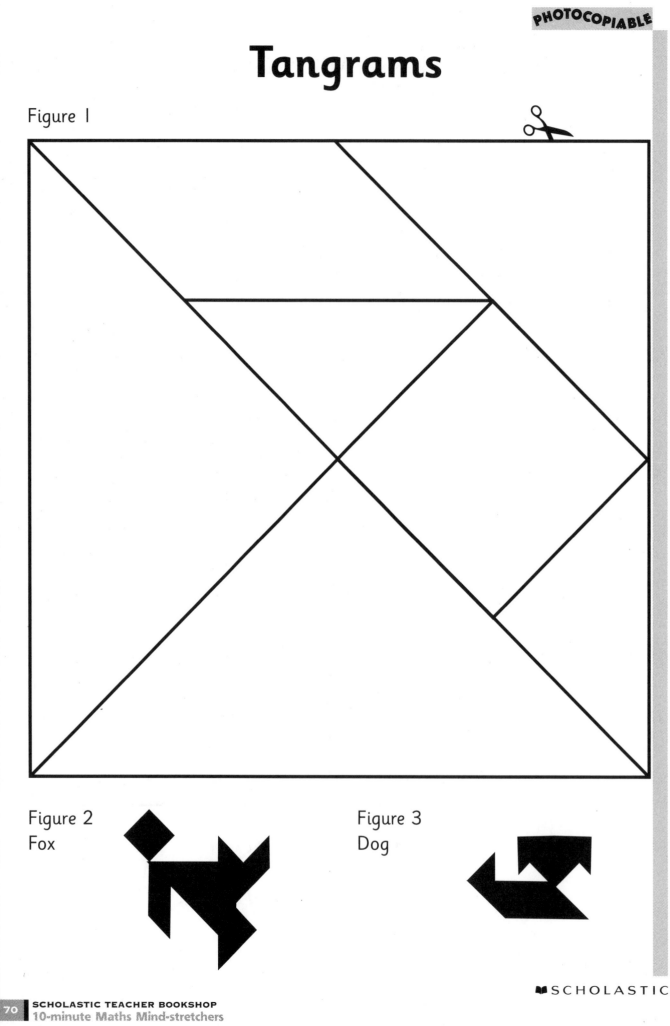

Figure 2
Fox

Figure 3
Dog

►SCHOLASTIC

Origami cup

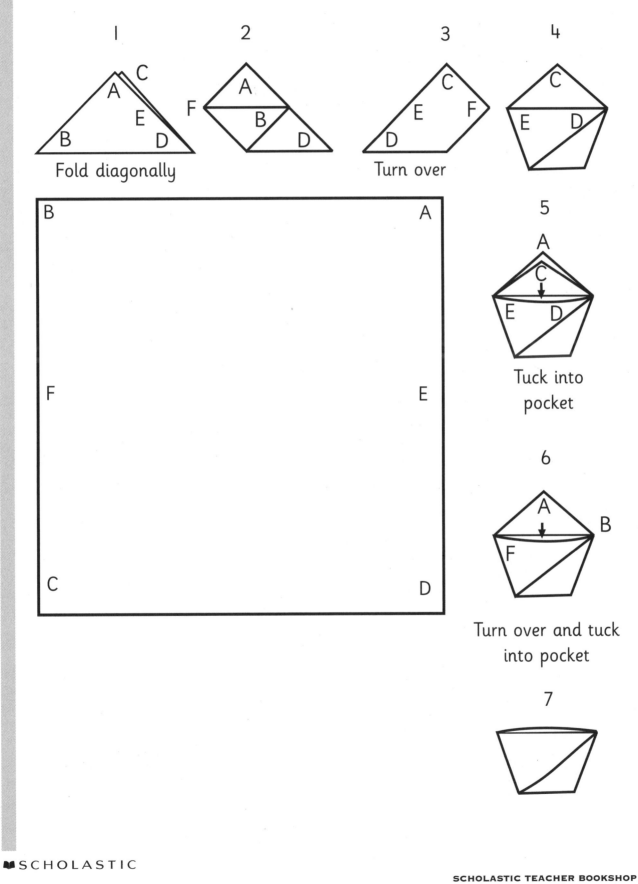

1

C
A
E
B D

Fold diagonally

2

A
F B
D

3

C
E F
D

Turn over

4

C
E D

5

A
C
E D

Tuck into pocket

6

A
B
F

Turn over and tuck into pocket

7

B

F

C D
A E

Chapter Eight

HANDLING Data

1 Bar graph of household chores

Organising and handling data

Make a bar graph to show which household chores the children like the most (or mind the least!) Draw a graph similar to the 'Basic graph' (photocopiable page 79) on the board. On the horizontal axis write five chores, such as cooking, washing up, emptying the dishwasher, tidying their bedroom, cleaning the pet's bedding (or draw symbols to represent each chore if you like). Ask each child to choose his or her favourite chore from those listed, and write the children's names to form columns up the vertical axis for each category. Ask children to copy the information onto individual copies of the graph taken from page 79.

Question the children on the information shown in the graph, such as: *Which job is liked the most? Which job is liked the least? How many more like washing up than like emptying the dishwasher?*

Graphs will vary.

2 Line graph of class attendance

Organising and handling data

Make a line graph to show your class' attendance for one week.

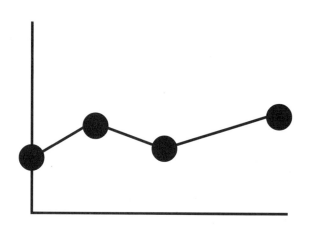

Use the 'Basic graph' worksheet (page 79). Along the horizontal axis write the days of the school week. Up the vertical axis write the total number of children in your class.

Record daily attendance by finding the line for the day of the week across the bottom of the graph, and the number on the vertical axis that represents the number in class that day. Mark the point where these lines intersect with a dot.

At the end of the week, draw a line connecting the dots and describe what this line looks like. What does the line tell you about class attendance?

Make sure children's graphs are accurate.

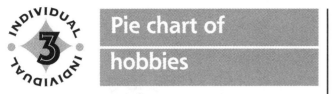

3 INDIVIDUAL

Pie chart of hobbies

Organising and handling data

Look at the pie chart in figure 1 on photocopiable page 80. In a pie chart, each section represents a part, or percentage, of the whole. The entire circle represents one whole or 100%.

Fill in the pie chart to show how much of Dilip's free time he spends on different hobbies. Use the following information:

20% cricket

30% playing drums

30% computer games

10% watching TV

10% drawing.

Now draw a pie chart that shows how much of your free time you spend on your different hobbies.

Check children's graphs. Individual graphs will vary.

4 INDIVIDUAL

Comparing two hobbies

Organising and interpreting data

Look at the graph in figure 2 on photocopiable page 80. What is the label on the horizontal axis? What is the label on the vertical axis? Write one or two sentences describing what the graph shows about Ralph and Anya.

The axes represent playing football and drawing; Ralph likes to draw more than he likes to play football, and more than Anya likes drawing; Anya likes to play football more than she likes to draw, and more than Ralph likes playing football.

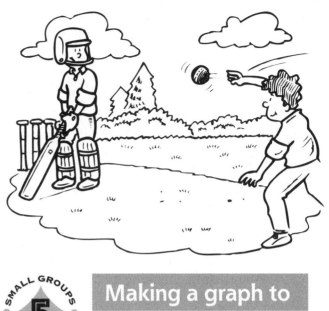

5 SMALL GROUPS

Making a graph to compare two hobbies

Organising and presenting data

Make a graph of your own like the one in figure 2 on photocopiable page 80. Label your graph with two activities and a measuring system such as 'High interest' and 'Low interest', or 'Dislike' and 'Like very much'.

Ask the other members in your group to mark the graph to show how they would rate themselves on the two activities. Write one or two sentences describing what your graph shows.

Graphs and descriptions will vary.

6 SMALL GROUPS

Taking a survey

Organising and presenting data

You can conduct surveys to gather information about the likes and dislikes of people in your school. Work with the members in your group to decide what information you would like to gather. Write three questions to ask in your survey. You could ask questions about sport, hobbies, foods, music and so on.

Find ten people who you can survey and then draw a graph to show your results.

Survey questions and answers will vary. Check children's graphs.

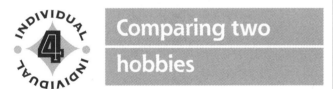

LITERACY

Reading speed (Literacy)

Ask children to label three columns along the horizontal axis of the graph on photocopiable page 79: 'Book 1', 'Book 2', 'Book 3'. Label the vertical axis from 1 to 200 in multiples of 10. Time children as they read for one minute. Ask them to count the number of words they read (while still understanding the story) and record the result in the first column of their graphs. Repeat with a second and a third book for each child until they have recorded the number of words read in one minute for three books. What title would they suggest for their graphs?

Discuss factors that make a difference in reading speed, such as difficulty of material, number and size of pictures, size of print, number of words on a page and so on.

7 Using coordinates to plot points on a grid

PAIRS PAIRS PAIRS PAIRS

Organising and presenting data

Look at the grid in figure 1 on the 'Grids and charts' worksheet (photocopiable page 81). The numbers across the bottom identify positions along the horizontal axis. The letters up the left side identify positions on the vertical axis. A number and a letter together can be used to identify the location of a point on the grid. The number and letter together are called the coordinates of the point.

Draw a dot on the grid to show the point named by each of these coordinates:

9e	8d	8f	7c
7g	6b	6h	5a
5l	4b	4h	3c
3g	2d	2f	1e

Connect the dots: what shape do you see? Compare your grid with your partner's. If any of your points are in different locations, check the coordinates to see which location is correct.

The given coordinates form a rhombus, or diamond.

8 Noughts and crosses

PAIRS PAIRS PAIRS PAIRS

Shape and space

Use the grid in figure 2 on the 'Grids and charts' worksheet (photocopiable page 81) to play a game of noughts and crosses. You and your partner take turns naming coordinates and drawing an 'X' or 'O' on the grid to mark the point named. The object of the game is to be the first one to get 3 Xs or Os in a row diagonally, vertically or horizontally.

Answers will vary.

9 Following and giving directions to draw pictures on the grid

PAIRS PAIRS PAIRS PAIRS

Handling data

Make two grids like the one in figure 1 on the 'Grids and charts' worksheet (photocopiable page 81), but label the spaces with numbers and letters instead of the lines. Create a picture or design on one of the grids by colouring in some or all of the squares.

Without showing your partner what you drew, give him or her the coordinates so that he or she can copy your picture or design onto the blank grid.

Pictures and designs will vary.

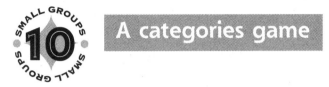

A categories game

Organising and presenting data

Use the chart in figure 2 on the 'Grids and charts' worksheet (photocopiable page 81) to play a categories game. In the horizontal spaces, write the letters 'R', 'M', 'T', and 'P'. In the spaces up the left write 'food', 'animal', 'clothing', 'sport'. Move across and up to find out what to write in each square. For example, in the square where the T column intersects with the *food* row, you must write a food that begins with T, such as *tacos* or *tomatoes*. Write two items in each square.

	R	M	T	P
Food			Tacos Tomatoes	
Animal				
Clothing				
Sport				

Compare your chart with the charts of the other members of your group. Score one point for each item you have that another group member has too; score two points for each item you have that others don't. The person with the greatest number of points wins.

Charts and scores will vary.

Always, often, sometimes, never

Organising and presenting data

Label the spaces along the horizontal axis of the chart in figure 2 on the 'Grids and charts' worksheet (photocopiable page 81): 'always', 'often', 'sometimes' and 'never'. Write the following activities along the vertical axis: do chores, study for tests, eat dessert, go to bed early. Decide how often you do each of these and put a tick mark in the appropriate box.

Compare your chart with your partner's. Describe how your charts are similar, and how they are different.

Charts will vary. Similarities and differences will vary.

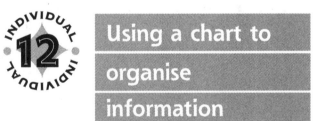

Using a chart to organise information

Organising and presenting data

Write the steps you would follow to find the answer to the question: *In which month do the greatest number of people in your class have their birthday? How would a chart be helpful?*

Steps will vary, but may include: 1 Write down the birthdays of everyone in the class. 2 List the months and make a tally mark next to the month for each birthday. 3 Count the tally marks to find which month has the greatest number. One possible response: a chart would be helpful for listing the months and making a tally.

MATHS

Hide the marker

Children can play this game in groups of two to four. Each player needs a copy of the grid in figure 1 on photocopiable page 81, and a small marker, such as a centimetre cube. One player hides his or her marker on a point of the grid, keeping the grid out of view of the other players. The other players take turns naming coordinates, trying to guess where the marker is located. If the guess is not correct, the 'hider' tells the player one direction to move (up, down, left or right) to get from the coordinates which were guessed towards the hidden marker. The other players move their own markers according to the clues the hider gives. The player who finds the coordinates of the hidden marker hides his or her marker for the next game.

* Label the horizontal lines (from the bottom to the top): 1st Street, 2nd Street, and 3rd Street.

* Label the vertical lines (from left to right): Hill Avenue, Lake Avenue and Park Avenue.

* Write 'N' (for north) at the top of the map, 'S' (for south) at the bottom, 'W' (for west) on the left side, and 'E' (for east) on the right side.

* Draw a school on the north-east corner of 3rd Street and Hill Avenue.

* Draw a park extending from Lake Avenue to Park Avenue and from 1st Street to 2nd Street.

* Draw a pond in the south-west corner of the park.

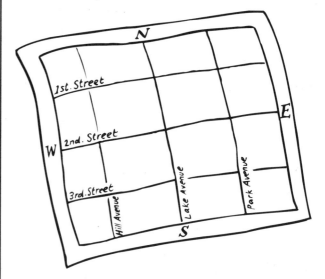

14 Finding locations on a map

Shape and space

Draw four places of interest on the map you made in the previous activity, such as tennis courts, a skating rink, a shopping centre, or the Town Hall. Take turns with your partner naming the location of each place by identifying streets, avenues or compass directions. Ask your partner to point out where it would be located if it were on his or her map.

Places drawn on maps will vary.

13 Making a map

Shape and space

Follow these directions to draw a map of a neighbourhood:

* Draw a grid with three horizontal lines and three vertical lines.

Creating map legends

Shape and space

Maps often have legends which explain what the symbols used on the map represent. Make a grid and draw a map of a city on it. Include five different symbols for things in the city such as schools, parks, train stations, libraries and so on. Include a legend that explains what each symbol means.

Maps will vary.

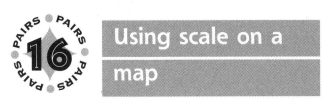

Using scale on a map

Shape and space

Maps are smaller than the areas they represent. A scale on a map explains how the measures on the map compare with the actual measures. For example, the scale on a map might be '1cm equals 10km'. If the distance between two towns on the map is 2cm, the actual distance is 20km. What would be the actual distance if the distance on the map was 0.5cm?

Work with your partner to draw a large rectangle to use as the outline of a map. Use a scale of '1 centimetre equals 10 kilometres' (1cm = 10km). You can think of a centimetre as being about equal to the width of your index finger. Label the towns of Riverville, Lakeville, Blueville and Redville on your map, using the following information:

★ *Riverville is 50km west of Lakeville.*

★ *Lakeville is 70km north of Redville.*

★ *Blueville is about half the distance between Riverville and Redville.*

Compare your map with another pair's map. Check that the distances between the towns are about the same. (The location of towns may be different, but the distances between them should be about the same.) If they're not, re-measure them.

Five kilometres.

Mapping the classroom

Shape and space

Ask children to work in small groups to draw maps of the classroom. Divide the room into sections and assign a section to each group (the size of each section will depend on the size of your classroom). Give each group a piece of graph paper, and help them determine a scale to use. For example, suggest that they use their feet to represent each centimetre of the graph paper. They can then use that reference measure to estimate the number of feet across the room, between objects and so on. Put the finished maps together to make a complete map of the room.

Maps will vary.

18 INDIVIDUAL

Estimating to compare distances between towns on a map

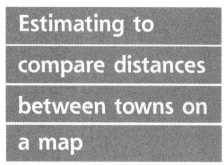

Estimating

Look at the map on photocopiable page 82. Estimate to compare distances between towns. Copy and complete this sentence at least three different ways:

_____ is about as far from _____

as _____ is from _____ .

Answers may vary, but could include: Poptown is about as far from Strawtown as Strawtown is from Globetown; Sandtown is about as far from Poptown as Looptown is from Globetown; Looptown is about as far from Rocktown as Rocktown is from Sandtown.

19 PAIRS

Designing a dream room

Shape and space

Design the room of your dreams. Decide on a scale, such as one centimetre equals one metre. Draw the objects in your room according to that scale. Colour and label the furniture in your dream room. Compare your picture with your partner's. Discuss ways the rooms are alike, and ways they are different.

Check children's drawings for consistent use of scale.

ACROSS THE CURRICULUM

GEOGRAPHY

Map collections

Collect a variety of maps, such as a street map of your town, road maps of counties, a political map of a country, a topographical map of a region, and so on. Display the maps for children to look at during free time. Ask them to compare two of the maps and write three ways in which the maps are alike, and three ways in which they are different.

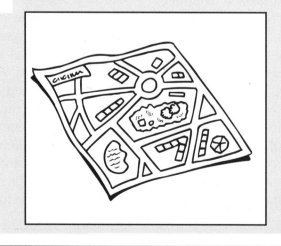

Visit from an architect

Ask a parent or another adult who is an architect to visit your class and discuss his or her job, emphasising how important maths is to the profession.

D & T

Basic graph

Graphs

Figure 1

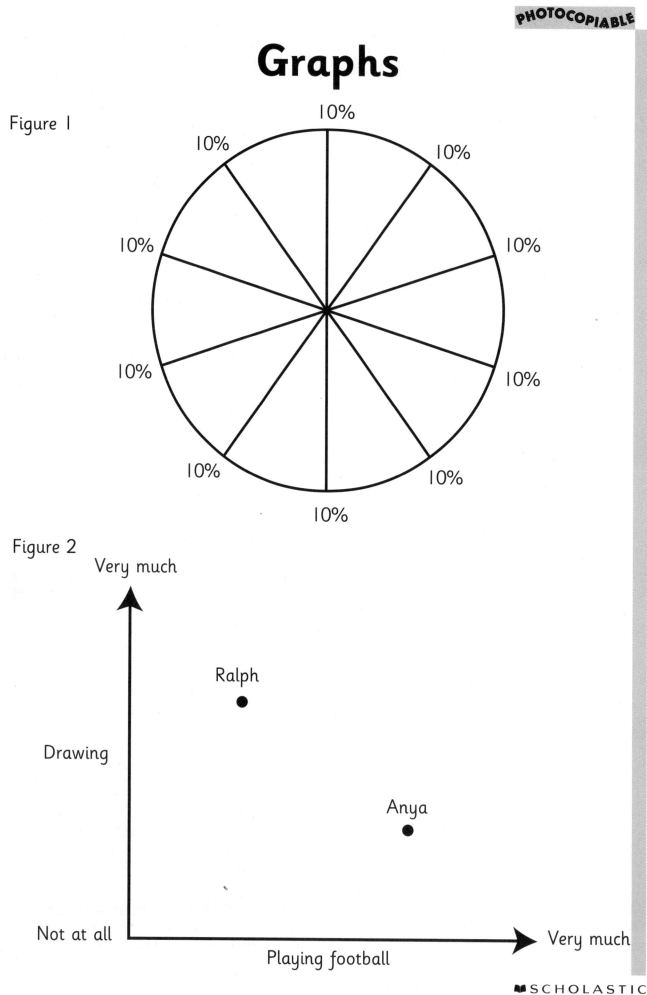

Figure 2

■SCHOLASTIC

Grids and charts

Figure 1

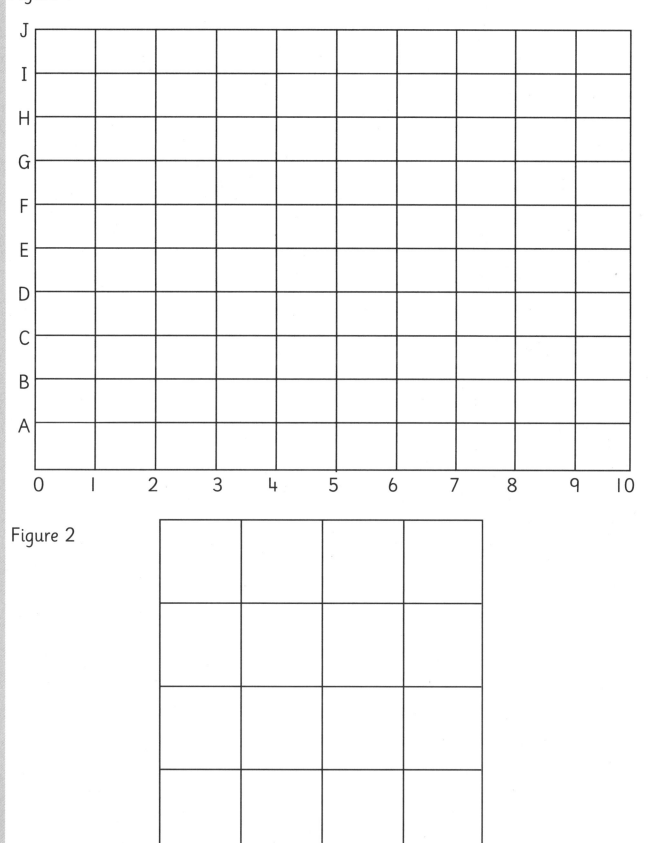

Figure 2

Maps and diagrams

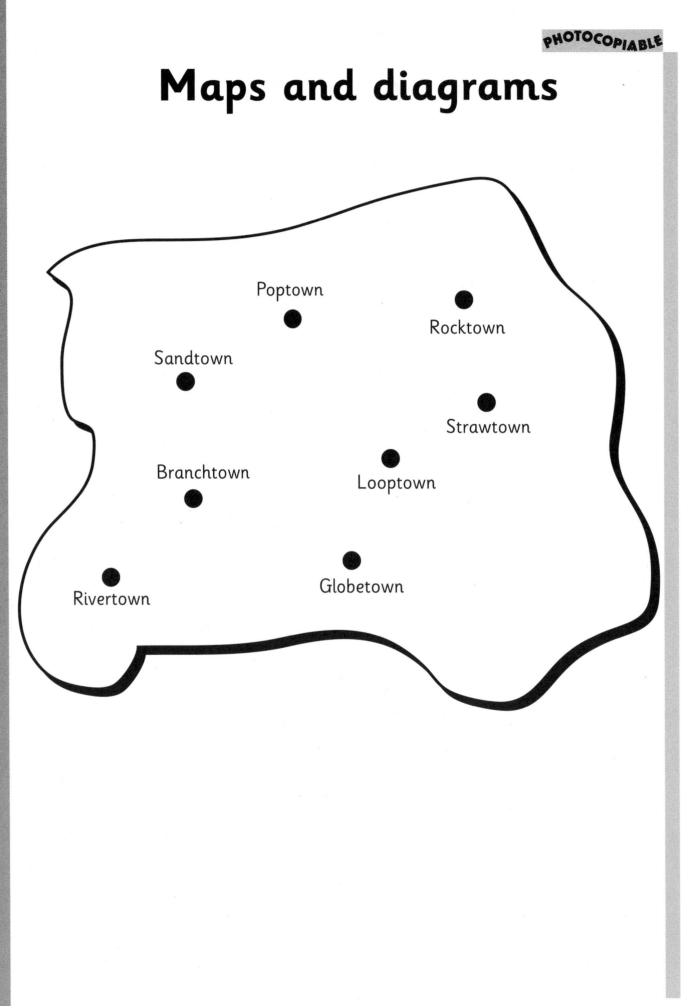

◄SCHOLASTIC

Chapter Nine

TIME and MONEY

1 INDIVIDUAL

Cycles in nature and measuring time

Measures

Long ago people used cycles in nature to tell the passing of time and seasons. Explain what time periods people measured using the Sun, the Moon and the seasons.

Sun – hours and days; Moon – days and months; seasons – years.

2 INDIVIDUAL

Days in a year

Measures

Would the number of days in a year be greater or less than 365 if each month had 30 days? Why does a year last for 365 days? Do some research to find out why we have a leap year every four years.

There would be less than 365 days; that's about how long it takes the Earth to complete one orbit around the Sun; the time it takes the Earth to orbit the Sun is closer to 365¹/₄ than 365 days, so every four years (a leap year) a day is added to adjust the calendar.

3 PAIRS

Time measured in twelfths

Measures

The number 12 is a special number when it comes to measuring time. Work with your partner to name at least two ways we measure time in twelfths or numbers that are divisible by 12. What other number do we use to measure time?

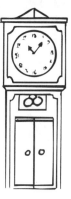

Answers may vary, but may include: there are 12 months in a year; days are measured by a multiple of 12 (24 hours in a day); seconds and minutes are measured in a multiple of 12 (60 seconds in a minute, 60 minutes in an hour); other numbers used to measure time may include: seven (days in a week); four (weeks in a month and quarters in an hour).

4 SMALL GROUPS

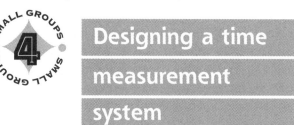

Designing a time measurement system

Measures

Suppose you could change the way time is measured. Work in a small group to write a short story explaining what changes you would make, and how the instruments we use to measure time, such as clocks and calendars, would be different.

Answers will vary.

Days in a month

Measures

Answer the following questions without looking at a calendar:

1   If the first day of the month is a Monday, what will be the dates of the other Mondays in the month?

2   If the first day of the month is Sunday, what will be the date of the first Saturday?

Write a calendar problem of your own. Swap problems with your partner and then later check each other's answers.

---

1 8th, 15th, 22nd, 29th; 2 the first Saturday wil be the 7th; problems will vary.

Comparing Mars years and Earth years

Measures

It takes Mars the equivalent of 687 Earth days to orbit the Sun. So, one year on Mars is 687 Earth days. About how many Earth years is 687 days?

Use a calculator to work out how many days old you are. How many years old would you be if Earth years were as long as Mars years?

A year on Mars is nearly twice as long as a year on Earth; the number of days old and your age in Mars years will vary depending on age, but your age in Mars years will be roughly half your age in Earth years. For example, a child who is ten years old is 3650 days old. That's a little more than five years in Mars years.

ACROSS THE CURRICULUM

SCIENCE

Make a sundial

Follow these directions to make a sundial. You may wish to have each small group make a sundial, or make one sundial together as a class.

★ place a length of dowel in the ground where there is sunlight all day long
★ draw a large circle around the dowel
★ mark and label the places on the circle that the shadow points towards at 9am, 10am, 11am, 12 noon, 1pm, 2pm and so on.

Check the sundial against a clock or watch each day, making adjustments until it is accurate.

Ask questions about the experience, such as: *When does the shadow increase in length? Decrease in length? Is there a time when the shadow nearly disappears?*

The further from noon the hour is, either earlier or later, the longer the shadow. The shadow nearly disappears at noon.

7 — Inventions of the past 200 years

Measures

Tortoises live longer than any other animal. Some tortoises live to be 200 years old! Suppose you were born 200 years ago. Work with your partner to list five things that would have been invented or discovered during your lifetime. You can use an almanac or encyclopedia to help.

Answers will vary, but may include: telephone, television, computer, radio, microwave, car, train or aeroplane.

8 — Shift work

Measures

Make a chart showing whether people in the following jobs work during the day, at night, or both: a cook, a teacher, a police officer, a secretary, a security guard, a bank clerk, a supermarket checkout assistant, a doctor, a firefighter, a postman.

   Compare your group's chart with another group's.

Answers may vary. One possible answer: daytime only – a teacher, a secretary, a bank clerk, a postman; night-time only – none; both – a cook, a supermarket checkout assistant, a doctor, firefighter.

9 — Birth dates

Measures

Peg and Greg are twins. Peg was born at 11:57pm on December 31, 1999. Greg was born seven minutes later. Draw a clock to show when each twin was born. What is Peg's birth date? What is Greg's birth date?

The clock for Peg should show 11:57pm, the clock for Greg 12:04am; Peg's birth date is 31st December 1999 (or 31/12/99) and Greg's birth date is 1st January 2000 (or 1/1/00).

10 — Making a timetable

Measures

The Elscot Eagles' football matches last 1 hour and 30 minutes. The coaches need to timetable five games on one pitch on a Saturday. All teams must be off the pitch by 5pm. Make a timetable to show when each of the five games could be played.

Answers will vary according to starting time and whether any time is allowed between games. One possible timetable, allowing 15 minutes between games: Game 1 – 8:30am, Game 2 – 10:15am, Game 3 – 12 noon, Game 4 – 1:45pm, Game 5 – 3:30pm.

Making a party timetable

Measures

Laura is having a party. She wants her guests to spend some time swimming, eating lunch and watching a movie. Everyone will be doing the same activity at the same time. The party starts at 11:00am and ends at 2pm.

Which activity should take the longest amount of time? The shortest? Write a timetable for Laura's party. Compare your timetable with your partner's. Should they be the same? Explain.

Answers will vary; timetables will vary. One possible timetable: movie from 11am–12:30pm; swim from 12:30pm–1:30pm; eat lunch from 1:30pm–2pm; timetables don't have to be the same.

Making an after-school timetable

Measures

On Fridays, Sonia's class always has a test. Their teacher asks them to study for about two hours during the week for the test. Sonia likes to play outside in the afternoons, and her bedtime is 8:30pm. She reads for 30 minutes every night.

Together with your partner, write an after-school timetable for Sonia that includes her study time, the other activities mentioned, and other things Sonia might do in the evenings.

Timetables will vary. One possible timetable:

Time	activity
after school until 6pm	play outside
6–6:30pm	eat dinner
6:30–7pm	study
7–8pm	do chores, talk on phone, get ready for bed and so on
8–8:30pm	read

24-hour clock

Measures

In the military, people use a 24-hour clock. Look at the chart in figure 1 on the 'Military time' worksheet

(photocopiable page 91). Midnight is 0000 hours, 1am is 0100 hours, 2am is 0200 hours, and so on.

Write numbers on a clock face the way you are used to seeing them. Now draw another clock face and on it write military times. When the hands on a military clock have gone around once, how many times have the hands on a 12-hour clock gone around?

Check the children's clocks – the numbers 1 to 12 should be written on one clock face, the numbers 0000 to 2300 should be written on the other. The hands on a 12-hour clock go around twice for every one time around on a military clock.

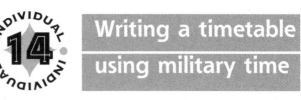

Writing a timetable using military time

Measures

Use the chart in figure 1 on the 'Military time' worksheet (photocopiable page 91) to rewrite your class timetable using military time.

Timetables will vary.

ACROSS THE CURRICULUM

CITIZENSHIP

World records

Provide children with the following information: a group of college students claim to have made the longest phone call in the world – they talked for over 720 hours!

Ask children to determine how many days that is (30 days). Working in small groups, using a world records book, to research three other unusual things people have done for a record amount of time. Make a class chart showing the record time for each event.

---

15 *PAIRS*

Estimating time

Measures

Luke groaned as he looked at the clock after he, his mother and two sisters finished a spaghetti dinner. His favourite TV show was starting, and it was his turn to wash up! Luke estimates it takes him ten seconds to wash each glass, 30 seconds to wash each dirty dish, and one minute to wash each dirty pan or serving bowl. He can wash all the silverware in five minutes. He needs to wash:

one spaghetti pan

one tomato sauce pan

one salad bowl

silverware, a glass, a salad plate, a dinner plate, and a dessert bowl per person.

About how many minutes of Luke's favourite programme will he miss?

Estimates may vary. He will miss about 15 minutes of his favourite programme. He spends about 100 seconds on each person's glass and dishes. For four people that's 400 seconds, or about seven minutes. He spends about three minutes doing the pans and serving bowl, and five minutes doing the silverware. The minutes add up: 7 + 3 + 5 = 15.

---

16 *INDIVIDUAL*

Objects used as money

Measures

Money has been in use for a very, very long time. People used shells, stones, clay, and other objects for money before it was made from metal and paper.

Suppose peanuts were used for money. If three peanuts are equal in value to a penny, how many peanuts would equal a five pence piece? Ten? Twenty? Would the value of 100 peanuts be greater than or less than 40p?

---

15 peanuts = 5p; 30 peanuts = 10p; 60 peanuts = 20p; the value of 100 peanuts would be 33$\frac{1}{3}$p – less than 40p.

Creating a money system

Measures

Work with your partner to invent a money system, using objects from nature in place of coins or notes. For example, five pebbles might equal one shell. It might cost two shells to buy a drink. Write or draw a description of your money system showing how many of one object it takes to equal the value of another object. Show the price of three things, such as an ice cream cone, a cinema ticket, and a pair of jeans using your system.

Answers will vary. Children may set up a system in which value is related to size – the smaller the object, the lesser the value, and it may take several small objects to equal the value of a large object.

Matching coins

Measures

Use the 'Coin cards' (photocopiable page 92) to find the card or combination of cards for each of these problems. Sets of play euro coins to support this activity are available from Learning Resources (website: www.learningresources.com/uk; tel 01553 762276).

1 Find three coins that can be combined to equal 40c.

2 Find the least number of coins that can make up €5.

3 Find the greatest number of coins that can make up €1.

4 What four coins can be combined to total 50c?

---

1 Cards showing 20c, 20c, 10c; 2 The least number of coins to make up €5 will be the cards showing €2, €2 and €1; 3 The greatest number of coins to make up €1 from those on the worksheet will be three cards showing 1c, a card showing 2c, three cards showing 5c, two 10c cards and three 20c cards; 4 Answers will vary.

Coin game

Measures

You will each need your own set of 'Coin cards' (photocopiable page 92) to play the coin game. Each player shuffles his or her set of cards and places the cards in a stack, face down. Next, each player turns over the top card on his or her stack at the same time. The player whose card shows the greatest amount of money records that amount. The first player to reach €5.00 or more wins. Sets of play euro coins to support this activity are available from Learning Resources (via their website: www.learningresources.com/uk; tel 01553 762276).

Answers will vary.

ACROSS THE CURRICULUM

GEOGRAPHY

Money around the world

Ask children to find out what money is called in other countries, and compare its value to the British pound or the Euro. (Foreign currency exchange rates can be found on the financial pages of most newspapers.) If possible, bring in, or get the children to bring in, currency from other countries. Display the money and discuss similarities and differences.

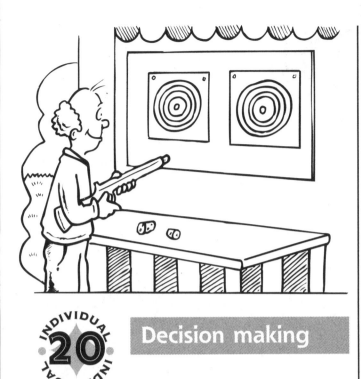

each week, how long will it take her to save enough money to buy the CD? How long would it take her to save enough money to buy two CDs at the same price?

Nine days: over a Monday to Sunday period, Anne can earn £10: £3 pocket-money + £1 per day for seven days of dog walking. In the next two days, Monday and Tuesday, she will earn the additional £5 she needs – £3 pocket-money + £1 per day for two days of dog walking. It will take her 3 weeks to save the £30 to buy two CDs.

20 Decision making

Measures

Pav and Ajay are going to spend two hours at an amusement park. The park has food booths, game booths and rides. They each have £10 to spend. They each have a choice of paying 74p for each ride, or paying £10 for an unlimited number of rides. What do Pav and Ajay need to think about in order to make their decision?

Answers will vary. One possible response: Pav and Ajay need to think about whether they want to spend any money on food or games, and whether they have enough time to make the unlimited number of rides worthwhile. In considering time they should think about the number of rides they want to go on, how long they would have to wait in line for each ride, and how long each ride may take.

21 Determining savings using a calendar

Measures

Anne wants to save her money for a CD that costs £15.00. She has £3 pocket-money, which she gets on Monday each week. She also has a job walking her neighbour's dog once a day. Each time she walks the dog she receives £1. If she saves all of her money

22 Making a purchase

Measures

Look at what is for sale in the shop windows of the shopping centre on photocopiable page 93. Suppose you won a £50 gift voucher.

1   What are two different ways you could spend all or some of the £50 at the Shopping centre? Write what you would buy and the cost.

2   How much more money would you need to buy a pair of in-line skates? A doll's house kit?

---

1 Answers will vary, but the total cost in either answer should not exceed £50; 2 £9.95 more for skates; £79.98 more for the doll's house kit.

Measures

The shops in the shopping centre on photocopiable page 93 are advertising sales. Make a table with a partner to compare the regular prices with the sale prices. Use the following information:

★ *Each CD is regularly priced at £12.99.*

★ *Everything in the sporting goods store is half the regular price.*

★ *Doll's houses usually sell for £89.98, and dolls are £10.00 each.*

★ *Paperback books are normally £4.85.*

Which store shouldn't be advertising its prices as sale prices?

Delia's Doll's Houses shouldn't be advertising sale prices. The correct answers are:

	Sale price	Regular price
Two CDs	£21.00	£25.98
Running shoes	£25.75	£51.50
In-line skates	£59.95	£119.90
Helmet	£19.99	£39.98
Doll's house and four dolls	£129.98	£129.98
Three books	£12.00	£14.55

Measures

Suppose you are a cashier. Explain what change you would give in the following situations, using mental maths when you can:

1 *£1 for a 15p item.*

2 *£5 for something that costs £3.99.*

3 *£10 for two items that cost £5.60 and £4.40.*

4 *£20 for three items that cost £6.25 each.*

Which ones did you solve using mental maths? Write four similar problems of your own and swap them with your partner.

---

1 85p; 2 £1.01; 3 no change; 4 £1.25.

Military time

12 midnight	= 0000 hours	12 noon	= 1200 hours
12.30am	= 0030 hours	12.30pm	= 1230 hours
1am	= 0100 hours	1pm	= 1300 hours
2am	= 0200 hours	2pm	= 1400 hours
3am	= 0300 hours	3pm	= 1500 hours
4am	= 0400 hours	4pm	= 1600 hours
5am	= 0500 hours	5pm	= 1700 hours
6am	= 0600 hours	6pm	= 1800 hours
7am	= 0700 hours	7pm	= 1900 hours
8am	= 0800 hours	8pm	= 2000 hours
9am	= 0900 hours	9pm	= 2100 hours
10am	= 1000 hours	10pm	= 2200 hours
11am	= 1100 hours	11pm	= 2300 hours

Coin cards

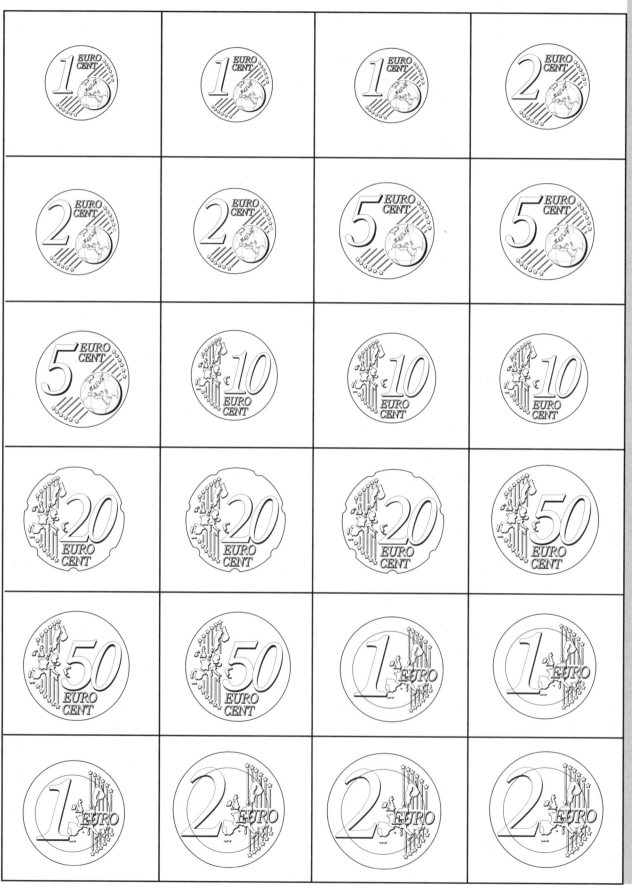

SCHOLASTIC

Shopping centre

Miguel's Music Sale!

Two CDs for only £21.00!

Spud's Sports Sale!

Everything in the shop is half price!

Running shoes now **£25.75**
In-line skates just **£59.95**
Helmets only **£19.99**

Delia's Doll's Houses

Hurry! Sale!

Just in... complete doll's house kit! Includes four dolls!

£129.98

Offer can't last!

Best Ever Books

Three paperbacks for £12.00!

Special Today only!

Chapter Ten

MEASUREMENT

Non-standard units of measure

Measures

Long ago, people measured length using the length of their own feet as the unit of measure. Since people have 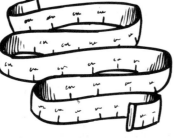 different sizes of feet, this caused some problems. Write a short story about a problem that occurred because people used their feet to measure. Explain how they solved the problem.

Stories will vary. Solutions will probably involve the creation or use of a standard unit of measure.

Using non-standard units of measure

Measures

Ask children to imagine they are living a long time ago – before rulers or yardsticks were used. Ask groups to decide what they would use to measure:

the length of a fish

the length of a cow from nose to tail

the height of a tree.

Groups should act out their solutions.

Answers will vary, but children should suggest using different units of measure for each: a larger unit of measure for the tree than for the cow, and in turn larger for the cow than for the fish.

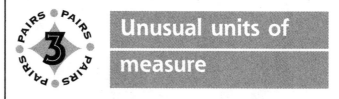

Unusual units of measure

Measures

Some unusual units are used for measuring very specific things. For example, the height of horses is measured in hands. One hand equals 4 inches (just over 10cm). The speed of a boat is measured in knots. One knot equals 1852m.

Working with your partner, make up a unit of measurement and name it. Explain what it would be equal to in length, capacity, speed or distance and what it would be used to measure (along with some sample measurements).

Answers will vary.

Weight versus capacity

4 INDIVIDUAL

Measures

Find out how cereal in a cereal box is measured. Why does a box of cereal sometimes not feel very full compared to another box the same size?

Cereal is measured by weight (grams). Two boxes the same size may not hold things that weigh different amounts but have a similar volume.

Accurate measures

5 SMALL GROUPS

Measures

Work with two or more children to brainstorm situations for which it is very important to have accurate measurements. Make a list of four situations.

Some possible answers: building houses, making clothing, giving medicine, planning a space shuttle flight.

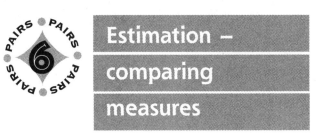

Estimation – comparing measures

6 PAIRS

Estimating

Use the 'Comparison cards' (photocopiable page 101) to make estimates about objects in your classroom. Then compare the objects to see if your estimates were correct.

Answers will vary.

ACROSS THE CURRICULUM

MATHS / SCIENCE / PSHE

Measure treasure hunt
Divide the class into small groups. Ask each group to find two things that:

★ weigh less than a textbook
★ are thicker than a piece of chalk
★ can hold more liquid than a coffee mug

Supermarket survey
Ask children to bring in lists of five or six food items and the units used to measure them. For example, drinks may be measured in litres; milk is measured in pints or litres; sugar is often measured in kilograms.

Measuring sound
Sounds we hear are measured in decibels. The softest sound a human can hear is about zero decibels. A normal speaking voice is about 60 decibels. Humans find it very uncomfortable to listen to a sound louder than 130 decibels. Think of four different sounds. Write them in order from softest to loudest. Copy the decibel scale shown on the right. Write your four sounds on the scale where you think they belong.

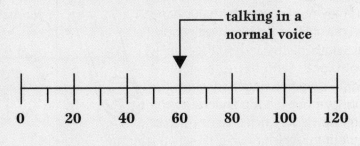

talking in a normal voice

0   20   40   60   80   100   120

Estimating length

Estimating

Estimate how many pennies placed in a row it would take to measure the length of a standard 15cm ruler. Check your estimate.

Estimates will vary. The ruler is eight pennies long.

Comparing circumferences

Measures

Describe a way to measure the distance around your wrist. Find an object with a greater circumference than your wrist, but smaller than the distance around your head. What shape is the object?

With a length of string or a tape measure; objects will vary; shape may be described as round, like a cylinder or like a sphere.

Reference measures

Measures

You may not know it, but your hand is a convenient ruler! For most people, the width of the index fingertip is about one centimetre, and most people have one finger on which the distance from the top knuckle to the fingertip is one inch. Use a ruler to find out for which of your fingers this is true. Use your 'built-in ruler' to find the length of your pencil and two other things at your desk. Record the answers both in inches and centimetres, then check your answers using a real ruler.

Answers will vary.

Inch, foot, yard

Measures

If 12 inches equal one foot, and three feet equal one yard, then how many inches equal one yard?

Find three things in the classroom that each measure about one yard. Find two things in the classroom that together measure about one yard. Record what you find.

There are 36 inches in a yard; lists will vary.

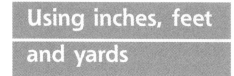

Using inches, feet and yards

Measures

Write two things that you would measure using inches, two things you would measure using feet and two things you would measure using yards.

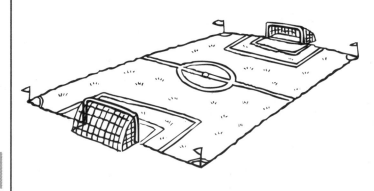

Answers may vary, but could include: inches – pencils, jeans (waist and seam measurements); feet – people's heights, rooms in homes; yards – football pitches, distances on playground.

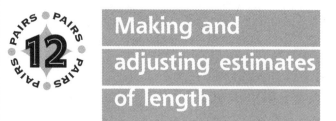

Making and adjusting estimates of length

Measures

Suppose ten children stood next to each other and held hands, stretching their arms as far as possible. Estimate how long the line they make would measure.

Work with a partner, stretch out your arms and ask them to measure the distance from fingertip to fingertip. How can you use this information to estimate the length of ten children's outstretched arms? What is your new estimate?

Which do you think is the better estimate, your first or your second estimate? Explain.

Estimates will vary; you can multiply the measure of your own outstretched arms by ten; estimates will vary; explanations will vary, but children should recognise that an estimate based on a known measure will probably be closer to the actual measure than an estimate based on an unknown measure.

ACROSS THE CURRICULUM

Go the distance

Make a ten-foot measuring rope by tying a knot at one end of a clothes-line. Tie additional knots at one-foot intervals, so that there are 11 knots in all. Cut off any extra clothes-line at each end. Children can count the knots to measure how far they can jump or travel in three hops.

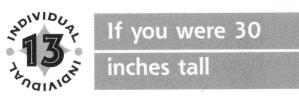
If you were 30 inches tall

Measures

The smallest dinosaur that ever lived was probably the Compsagnathus. It was only about 30 inches tall. Find something that is about that height. Record what you find.

Imagine you were 30 inches tall and think about the following:

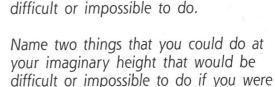

★ *List three things in the room that would be taller than you.*

★ *Name two things you often do at school that would be difficult or impossible to do.*

★ *Name two things that you could do at your imaginary height that would be difficult or impossible to do if you were your normal height.*

Answers will vary.

Perimeter

Measures

Work with your partner to cut out the squares in figure 1 on photocopiable page 102. Each square is 2cm long and 2cm wide. Make a figure with eight squares. (Make sure that any two squares that are placed next to each other touch along the entire side.)

Measure the distance around the figure you made by counting the outside edges of the squares (remember, the length of the side of each square is 2cm). The distance around a figure is called the perimeter.

Make a different figure using the eight squares. Does the perimeter change? Use the eight squares to make a figure with the greatest perimeter you can. What is the perimeter? Make a figure with the smallest perimeter you can. What is the perimeter?

Perimeters will vary; the perimeter may or may not change; the greatest perimeter of a figure is 36cm – a 1 × 8 rectangle; the smallest perimeter of a figure is 20cm – a 2 × 4 rectangle.

Estimating perimeter

Measures
Estimate the perimeter of a book. Now use a ruler to measure the exact perimeter. How does the actual answer compare with your estimate?

Estimate the perimeter of your desk, then measure the exact perimeter. Estimate, then measure the perimeter of your teacher's desk.

As objects become larger, does it get easier or harder to make an estimate? Explain.

Comparisons will vary; explanations will vary.

Estimating length

Measures
Estimate how many centimetres long your foot is. Estimate whether the length of the fingers and thumb on one hand, when added together, is more or less than the length of your foot. Measure to the nearest centimetre to check your estimates.

Estimates will vary. It is most likely that the sum of finger and thumb lengths will be greater than the length of the foot.

Identifying an object by length

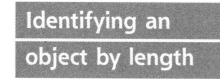

Measures
Look at the pictures on the 'Measuring objects' page (photocopiable page 103). Read the following clues to find out what the mystery object is:

The mystery object is longer than 10cm. It is not the longest object but it has the greatest perimeter. What is it?

The mystery object is the calculator. The perimeter is 42cm.

Describing objects using measurements

Measures
Using the objects on the 'Measuring objects' page (photocopiable page 103), make up your own measurement mystery. Give your partner the clues and ask him or her to guess your mystery object.

Clues and mystery objects will vary.

Area

Shape and space
When you find out how much space a figure covers, you are finding the area of the figure. It is measured in square units. Use all 25 of the 2cm squares in figure 1 on photocopiable page 102 to make three separate figures. What is the area of each figure? Use all 25 squares to make three more figures. Trace around the outside of your figures, record each area, and colour them.

Figures will vary, but the area of each figure will be 50 square centimetres.

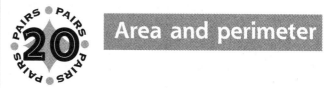

Area and perimeter

Shape and space
Work with a partner to make different figures using the 2cm squares in figure 1 (photocopiable page 102). Record the perimeter and area of the figures you make in the chart in figure 2.

Write a letter to a friend telling what you know about perimeter and area.

Charts will vary; letters will vary, but may include: figures with the same area can have different perimeters; to find the area of a rectangle, you can multiply the length by the width; to find the perimeter, you add the lengths of all the sides of a figure.

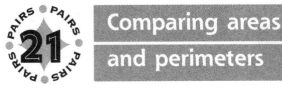

21 Comparing areas and perimeters

Shape and space

Look at the shapes on photocopiable page 104. Predict which ones have the same area and which ones have the same perimeter. Use the grid at the top of the page to check your predictions (each square of the grid is one square centimetre). Make a table like the one in figure 2 (photocopiable page 102) to record the measures.

Predictions will vary. Shapes A, B, D, E, and F all have areas of 12cm²; shapes A, C and D all have perimeters of 16cm.

Figure	Area (cm²)	Perimeter (cm)
a	12	16
b	12	14
c	16	16
d	12	16
e	12	26
f	12	18

22 Metre tape measures

Measures

Put two 100cm lengths of masking tape together, sticky sides facing each other (make enough for the class to work in pairs). Ask the children to use a ruler to mark each centimetre from 1 to 100 (starting from the left end), labelling multiples of five (5, 10, 15, 20 and so on). They can colour alternating sections of ten centimetres (from 0 to 10cm, 11 to 20 cm, 21 to 30 cm and so on) different colours to make it easier to read the tape measure.

Children should use their tape measures to find items in the room that are about one metre in length. Then point out objects in the classroom and ask children to estimate whether they are greater than or less than one metre in length. Choose different children to use their tape measures to check the accuracy of the estimates.

Objects chosen and estimates will vary.

23 Comparing size

Measures

Close your eyes and picture a jar filled with small marbles. Now picture a jar the same size filled with large marbles. Open your eyes and draw a picture of what you saw. Which jar do you think holds more marbles? Explain why you think this.

Check children's drawings. They should recognise that the same size jar will hold a greater number of small objects than large objects.

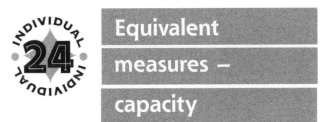

24 Equivalent measures – capacity

Shape and space

Copy the table below. Use the following information to complete it:

25ml = 1 fl oz; 150ml = ¼ pint or 6 fl oz; 300ml = ½ pint or 12 fl oz; 600ml = 1 pint or 24 fl oz; 1 litre = 1¾ pints or 40 fl. oz).

50ml = 2 fluid ounces (fl oz)

Table of measures

Unit of measure	Imperial equivalent
25ml	
150ml	
300ml	
	1 pint (20 fl oz)
1 litre (1000ml)	

25 Equivalent measures – weight

Measures

Copy the table below. Use the following information to complete it:

approximately 100–125g = 4 ounces (oz);
Imperial pound (lb) = 16oz

TABLE OF MEASURES

Unit of measure	Imperial equivalent
50g	
	1lb
225g	
1kg	

Accept reasonable answers. 50g = 2 oz;
225g = ½lb or 8 oz; 450g = 1lb or 16 oz;
1kg = 2lb 4oz.

26 Measuring weight

Measures

At a grocery store, a sign says that apples are 99p a kilo. If you didn't have a scale, how could you tell whether four apples are less than, more than, or equal to one kilo? Discuss it with your partner. Write or draw your answer.

Answers may vary. One possible answer: find
something that weighs 1kg, such as a bag of
sugar or flour, and compare the weight of that
object to the weight of the apples.

27 Temperature

Measures

Imagine you are standing outside, looking at a thermometer showing the temperature as 7°C. Would the air feel cold, cool, warm or hot?

Draw pictures of the ways you would dress to be outside in each of the following temperatures:

15°C   32°C   0°C   24°C

Air at 7°C would feel cold. Check children's
drawings: they should show clothing appropriate
for cool weather (15°C); hot weather (32°C); cold
weather (0°C); warm weather (24°C).

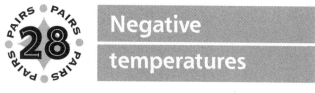

28 Negative temperatures

Measures

On a very cold winter morning the thermometer showed that the temperature was −2°C. In the early afternoon the thermometer showed that the temperature was 3°C. How many degrees did the temperature rise? Draw a picture of a thermometer to help you find the answer.

The temperature rose by 5°C.

Comparison cards

1	4
———————	———————
is as tall as	is wider than
———————	———————
2	5
———————	———————
is shorter than	is the same height as
———————	———————
3	
———————	
is the same length as	
———————	

Measuring squares

Figure 1

Figure 2

Number of squares used	Area	Perimeter

PHOTOCOPIABLE

Measuring objects

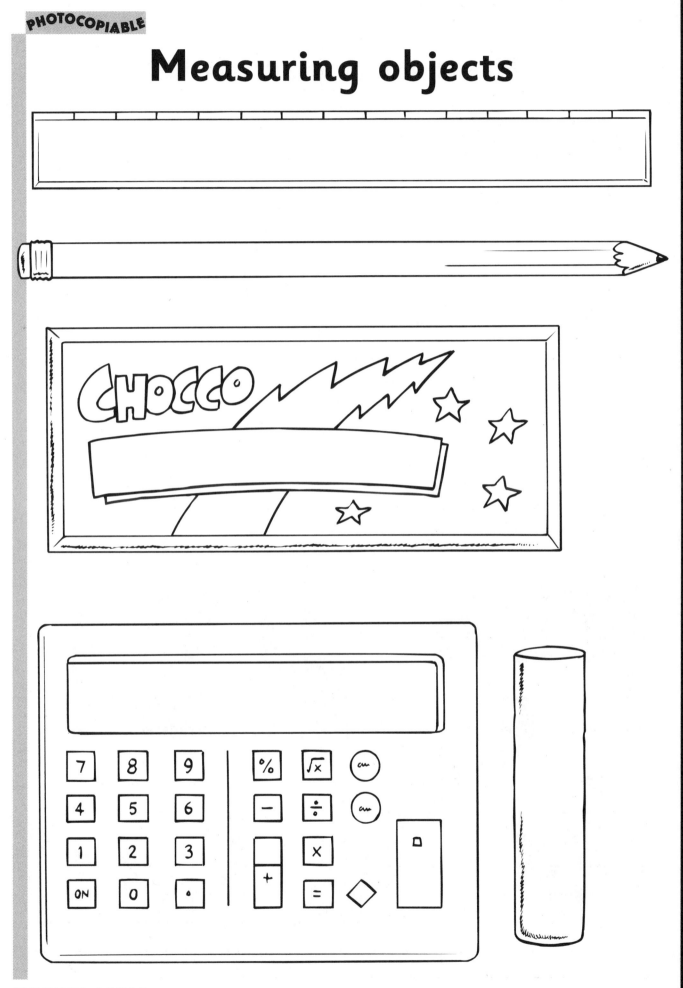

Area and perimeter

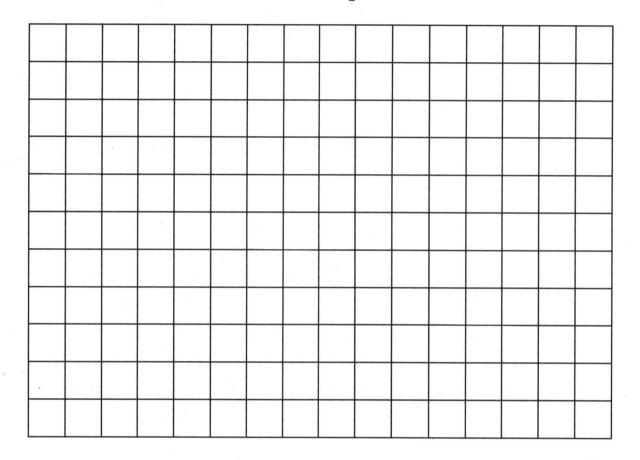

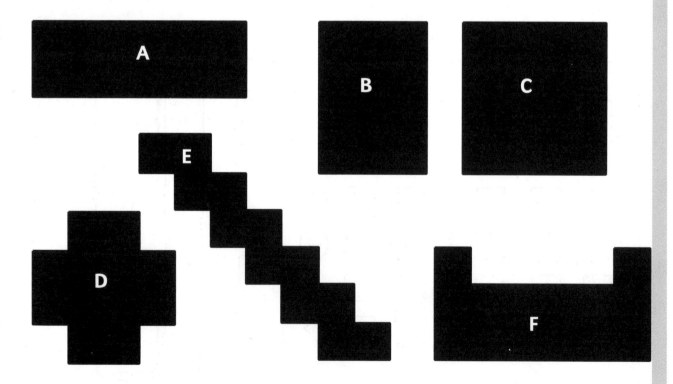

◢SCHOLASTIC

FRACTIONS, DECIMALS, PERCENTAGES solving

1 Designing a one-minute workout

Problems involving 'real life'
Sometimes we feel as though we've been sitting still too long, and we need to exercise our muscles. Work with a partner to create a one-minute exercise workout. Make a plan to do at least three different exercises in that time. Estimate how many times you think each exercise can be done. Take turns timing and performing the exercise routine until you get the time down to one minute.

Exercise routines will vary.

2 Solving a number puzzle

Problems involving 'real life'
Complete the circle puzzle on photocopiable page 111, following these directions:

1 Write a number in each of the empty sections.
2 Each number must be placed so that it is opposite a number, which is double or half its value.
3 The sum of all the numbers in the puzzle should be 36.

Compare your puzzle with your partner's. Make up a puzzle like this and swap with your partner to solve.

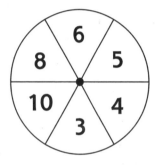

Chapter Eleven

3 Magic square

Problems involving 'real life'
In a magic square, numbers are placed so that you get the same sum whether you add the numbers in a column, across a row, or diagonally from corner to corner. Complete the magic square on photocopiable page 112 so that there is a number from one to nine in each box. Each number can be used only once.

6	1	8
7	5	3
2	9	4

Creating a magic square

Problems involving 'real life'

Create your own magic square, writing the numbers from 1 to 9 in different boxes in the square. Remember that each number can be used only once. Explain how you decided where to place the numbers.

Write in three numbers from your magic square in their correct places on an empty magic square. Swap with your partner and solve.

Magic squares will vary; the methods children use to place numbers in the magic square will vary, but may include: guess and check; writing down all the combinations of three numbers less than ten that have a sum of 15 and using that list to place the numbers in the square.

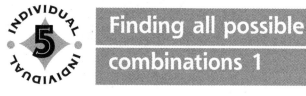

Finding all possible combinations 1

Problems involving 'real life'

James, Patrick and Joseph went on a picnic, each taking some food with them. Between them, they had peanut butter sandwiches, cheese sandwiches, bananas and apples, and bags of crisps and tortilla chips. Draw a picture to show the different lunches they could make if, for each lunch, they took one sandwich, a piece of fruit and one bag of crisps or chips.

Check children's drawings. They could make 2 × 2 × 2, or eight different lunches.

Finding all possible combinations 2

Problems involving 'real life'

For this problem you need five different coloured slips of paper. Work with your partner to find how many ways you can arrange the slips of paper in two groups. A group can have one or more slips of paper in it. Draw pictures or make a table to record the different groups.

There are 15 different ways to arrange the slips of paper into two groups - five ways to have one slip of paper in one group and four slips in the other; and ten ways to have two slips in one group and three slips in the other).

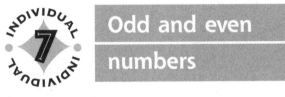

both even, your age will be an even number. If one of the years is odd and one even, your age will be an odd number.

Odd and even numbers

INDIVIDUAL 7 INDIVIDUAL

Properties of number

If the year you were born in was an even-numbered year, and this year is an even-numbered year, will your age on your birthday this year be an odd or an even number?

If the year you were born in was an odd-numbered year, and this year is an odd-numbered year, will your age on your birthday this year be an odd or an even number?

Will your age be an odd or even number if one of the years is an odd number, and one is an even number?

Write a rule you can use for any year to tell whether your age on your birthday that year will be an even or odd number.

Even; even; odd. Possible rule: if your birth year and the current calendar year are both odd or

both even, your age will be an even number. If one of the years is odd and one even, your age will be an odd number.

Measuring precipitation

INDIVIDUAL 8 INDIVIDUAL

Measures

Precipitation is the amount of water that falls on the Earth. It takes about ten centimetres of snow to equal the same amount of water as one centimetre of rain.

If an area of land receives 30 centimetres of precipitation in a year, what are five different combinations of snow and rain it could have received?

Answers will vary, but may include: 29cm rain/ 10cm snow; 28cm rain/20cm snow; 27cm rain/ 30cm snow; 26cm rain/40cm snow; 25cm rain/ 50cm snow.

ACROSS THE CURRICULUM

LITERACY

Logical scenes

Read the following problem about a cat, a mouse, and some cheese out loud, and then ask the children to act out the solution.

A man had to take a cat, a mouse and some cheese across a river. His boat was so small, though, he could only take one of them across at a time. He couldn't leave the cat alone with the mouse, or the mouse alone with the cheese. How could the man get all three across the river safely? How many trips across the river would he have to make?

The man could take the mouse across first, leaving it on the opposite shore, then returning to pick up the cheese. When he got the cheese across he'd need to take the mouse back. Then he could pick up the cat and take it across, leaving the mouse behind. Finally, he could go back for the mouse and bring it across again. It would take him seven trips across the river.

INDIVIDUAL 9

Calculating time

Measures

At one theme park, it takes four minutes to climb to the top of the water slide, and one minute to slide down it. Once they've landed in the pool below, people usually spend between 30 seconds and two minutes splashing around before getting out.

What is the greatest number of trips up and down the slide someone could make in 30 minutes? In real life, what other factors might you need to take into consideration?

A trip up and down the slide takes a minimum of five minutes and 30 seconds, up to a maximum of seven minutes. The greatest number of trips someone could make in half an hour is five: 5 × 5 minutes 30 seconds = 27 minutes 30 seconds. Other factors to take into consideration will vary, but may include: whether or not there is a queue, or whether you stop to talk or swim).

ACROSS THE CURRICULUM

GEOGRAPHY

World shopping trip (Geography)

Ask the children to imagine they can visit anywhere in the world, and they have £5000 to spend. They must use the money to buy a plane ticket, plus cover the cost of accommodation and meals for a week. Children can research the cost of the trip and make up a budget for it. In their budgets, they can record how much their accommodation costs each night, the price of the plane ticket, and the cost of each meal.

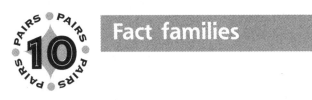

PAIRS **10**

Fact families

Properties of number

Number fact families are made up of two or three numbers that together make at least two related number sentences. For example, 4, 25 and 100 are a fact family. The numbers can be used to make the following number sentences:

$25 \times 4 = 100; 4 \times 25 = 100;$

$100 \div 4 = 25; 100 \div 25 = 4$

Choose two fact families and write the related sentences for each.

Write the numbers belonging to the two fact families in a mixed up order and swap papers with your partner. Separate the numbers into the two fact families and write their related sentences. Check each other's work.

Fact families and related sentences will vary.

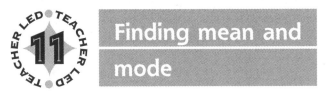

TEACHER LED **11**

Finding mean and mode

Handling data

Each child needs a penny for this activity. Follow the steps below to find the mode and mean for the number of years the children's pennies have been in circulation. Explain that you find the mode by

identifying the value that occurs most often. Explain that you find the arithmetic average, or mean, of a set of numbers by adding them and then dividing the sum by the number in the set. Both the mode and the mean are ways to describe an average value of a set of data.

1 *Ask each child to find the number of years his or her penny has been in circulation by subtracting the year shown on the penny from the current year.*

2 *Make a graph of the results. Ask children to work out the most common number of years pennies have been in circulation. This is the mode.*

3 *Add the number of years the pennies have been in circulation together and divide by the total number of pennies. This is the mean.*

Compare the mode and the mean.

Mode and mean will vary with pennies used. The mean calculated may or may not have the same value as the mode, and may or may not be a whole number.

12 INDIVIDUAL
Writing a plan for finding an average

Handling data

Ask children to write a plan to find the average number of brothers and sisters in the whole class.

Plans should include a way to collect the data. Discuss ways of finding either the arithmetic mean – adding the numbers of siblings and dividing by the number of children in the class, or the mode – identifying the number of siblings of children in the class.

13 PAIRS
Probability

Handling data

Work with your partner to solve this problem.
It was dark in the morning when Jack was getting dressed. He reached into his sock drawer, where he kept ten pairs of white socks and five pairs of black socks (of course, they weren't together in pairs!) What is the fewest number of socks Jesse would need to pull out of his drawer before he could be sure he'd have two matching socks? Explain your answer.

Three socks. The first sock he pulls out would be either white or black; the second sock might match the first, or it might not; the third sock will match the first sock, the second sock, both socks, or neither sock – but in that case, the first two socks would have to match.

14 INDIVIDUAL
Probability with number cards

Handling data

Madge and Norman were playing with a set of cards numbered 1 to 24. 'I'll bet I can pull a number greater than 20 out of this set of cards,' said Madge. 'I have a better chance of pulling out an even number than you do of pulling out a number greater than 20,' claimed Norman. Was he right? Can you explain why or why not?

Norman was right. There are four cards greater than 20 in the deck of cards, so Madge has four chances out of 24 cards, or a one in six chance of pulling a number greater than 20. There are 12 cards with an even number, so Norman has 12 chances out of 24 cards to pull an even-numbered card from the deck, or a one in two chance.

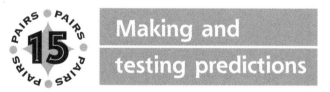

15 PAIRS
Making and testing predictions

Handling data

Predict which vowel is used most often in our language. To check your prediction, count 100 words from one or more pages in a book. Work with your partner to make a tally of the number of times each vowel occurs in those 100 words. Which vowel occurs most often?

Repeat the activity for another set of 100 words. Does your answer change? How did your answer compare to your prediction?

Predictions will vary; results will vary, but will probably show that 'e' is the most common vowel; comparisons will vary.

Number puzzles

Circle puzzle

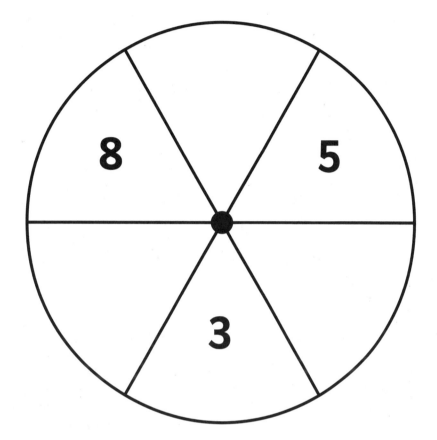

◄SCHOLASTIC

Number puzzles

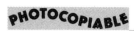

Magic square

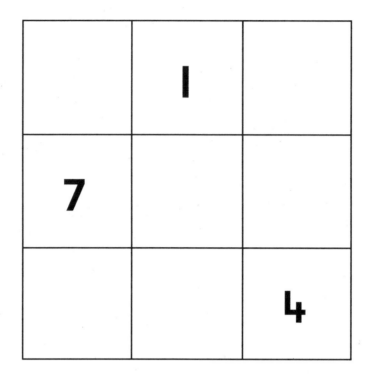